THE REPUBLIC OF TECHNOLOGY

Books by Daniel J. Boorstin

The Republic of Technology

•

The Americans: The Colonial Experience
The Americans: The National Experience
The Americans: The Democratic Experience

•

The Mysterious Science of the Law
The Lost World of Thomas Jefferson
The Genius of American Politics
America and the Image of Europe
The Image: A Guide to Pseudo-Events in America
The Decline of Radicalism
The Sociology of the Absurd
Democracy and Its Discontents
The Exploring Spirit
Portraits from The Americans
The Chicago History of American Civilization (27 vols.; editor)
An American Primer (editor)
American Civilization (editor)

•

For young readers

The Landmark History of the American People
Vol. I: From Plymouth to Appomattox
Vol. II: From Appomattox to the Moon

The Republic of Technology

Reflections on Our Future Community

DANIEL J. BOORSTIN

1817

HARPER & ROW, PUBLISHERS

NEW YORK, HAGERSTOWN

SAN FRANCISCO

LONDON

 For information address Harper & Row, Publishers, Inc., 10 East 53rd Street, New York, N.Y. 10022. Published simultaneously in Canada by Fitzhenry & Whiteside Limited, Toronto.

FIRST EDITION

Designed by Sidney Feinberg

Library of Congress Cataloging in Publication Data

Boorstin, Daniel Joseph.
The republic of technology.
Includes index.
1. Technology—Social aspects—United States.
2. United States—Intellectual life. I. Title.
T14.5.B66 301.24'3 77-11823
ISBN 0-06-010428-7

78 79 80 81 82 10 9 8 7 6 5 4 3 2 1

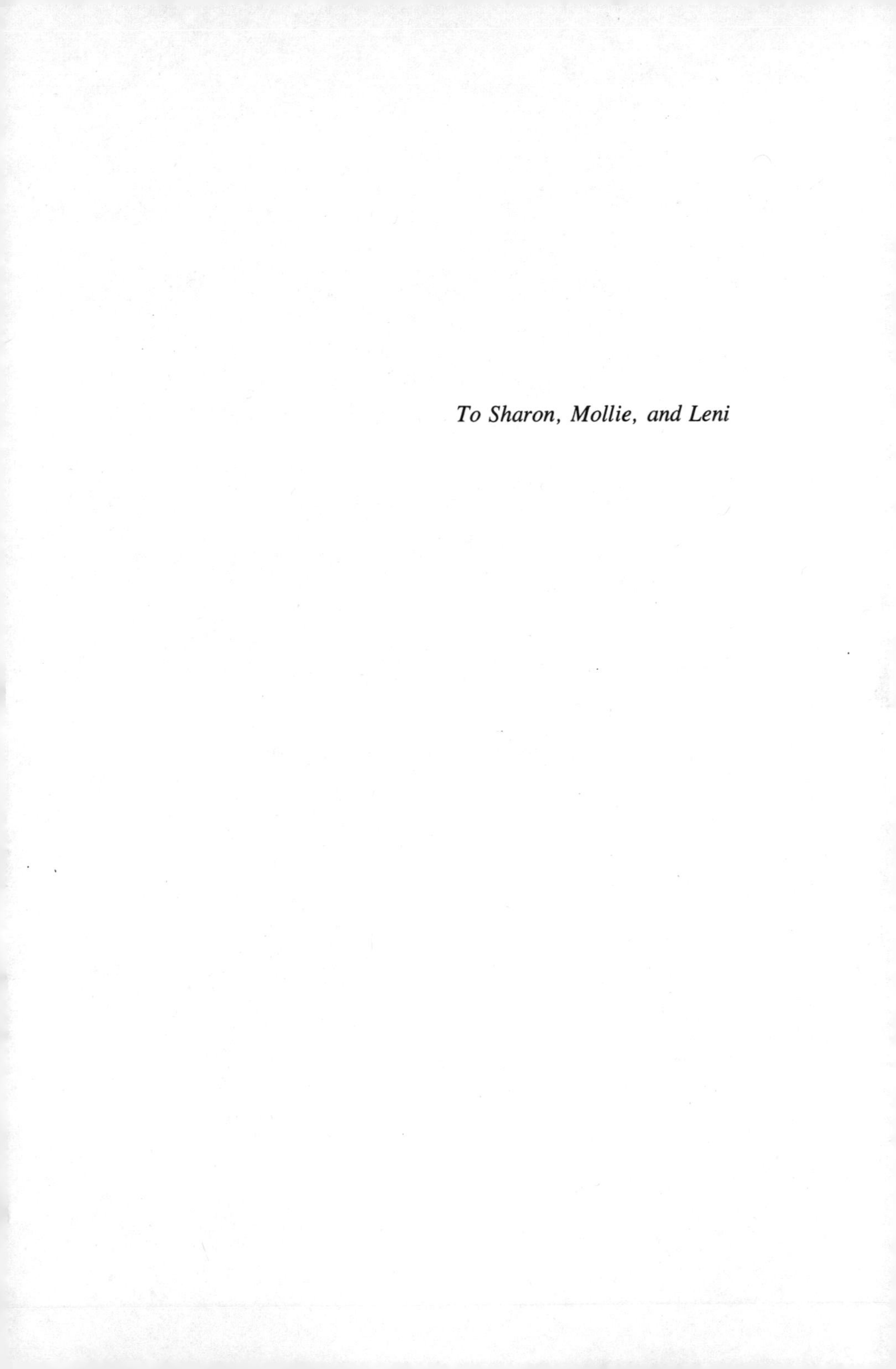

To Sharon, Mollie, and Leni

"We know what we are, but know
not what we may be."

Shakespeare

Contents

Foreword

Every year our nation becomes less peculiar. The very same new forces that have given a special character to life in America now every year make the lives and fortunes of people everywhere more like the lives and fortunes of people everywhere else. Science is the enlarging international pool of knowledge equally valid everywhere. Technology, a synonym for experiment, is a name for the applications of science, which transcend political boundaries, language, religion, and local tradition.

In the past, "culture" or "civilization" described the special qualities of life in one or another part of the earth. Love of the culture of one's place was called "patriotism." Its pathological (and more common) form—chauvinism or xenophobia—was distrust or hate of other cultures. Technology somehow leads us to conquer or ignore these passions. Though people around the world may not love one another any more than they did, yet their ways of life tend to become more and more alike. Such questions as What is the future of "the West"?—or of "the East"?—every year become more obsolete. In the long run there is only one question, and that concerns the future of humankind.

Recent wars, even more than earlier wars, have made opposing nations more alike. Wartime developments—radar, the search for nuclear fission and for techniques of delivering destructive rockets—led to an international competition (and collaboration) in technology that produced the atomic bomb and atomic energy, television, space travel, orbiting satellites for communication, and myriad other novelties. These have converged the cultures of nations, incidentally reducing differences between big nations and small.

Rapid technological change and obsolescence (both substantially modern phenomena) have reduced the difference between victor and vanquished—and have even given an ironic new advantage to a nation that has suffered massive destruction of its capital equipment. Rebuilding its industrial plant, with the aid of the victor, gives the defeated nation a prime opportunity to lift itself above the technological level of its conqueror.

These same overwhelming forces of technology that homogenize the culture of the human race have disrupted the international community of nations. Peoples who, because of poverty, or colonialism, or remoteness from metropolitan centers, have never had a "national" culture, now assert a specious nationalism. Larger national units no longer can so easily dominate the small. Miniature "nations," grandstanding on planet-wide television, demand equality with older, larger national units. While the United States has moved toward "one person, one vote," the United Nations—the whole international community—has moved toward "one nation, one vote."

And what is a nation? Of the more than ninety new nations admitted to the United Nations since 1945, half are less populous than North Carolina. Earlier nationalisms, unlike those of the later twentieth century, began in affirmation of long-established vernacular literatures, of coherent histories, of characteristic institutions, of distinctive religious, economic, or cultural interests, of traditional boundaries. But every day the word "nation" becomes more meaningless.

Yet the older nations, among whom we must count the United States, still live by a faith in their special traditions. One of our deepest national traditions is to be an international nation. As the most technologically advanced great nation in the late twentieth century we are a center from which radiate the forces that unify human experience. Ideology, tribalism, nationalism, the crusading spirit in religion, bigotry, censorship, racism, persecution, immigration and emigration restriction, tariffs, and chauvinism do interpose barriers. But these are only temporary. The converging powers of technology will eventually triumph. They triumph for a host of reasons, which we are only beginning to discover, and some of which we will explore in the following pages.

THE REPUBLIC OF TECHNOLOGY

1 The Republic of Technology

"An athlete of steel and iron with not a superfluous ounce of metal on it!" exclaimed William Dean Howells before the centerpiece of Philadelphia's International Exhibition celebrating our nation's hundredth birthday. He was inspired to these words by the gigantic 700-ton Corliss steam engine that towered over Machinery Hall. When President Ulysses S. Grant and Emperor Dom Pedro of Brazil pulled the levers on May 10, 1876, a festive crowd cheered as the engine set in motion a wonderful assortment of machines—pumping water, combing wool, spinning cotton, tearing hemp, printing newspapers, lithographing wallpaper, sewing cloth, folding envelopes, sawing logs, shaping wood, making shoes—8,000 machines spread over 13 acres.

Others, especially visitors from abroad, were troubled by this American spectacle, "I cannot say that I am in the slightest degree impressed," announced the English biologist Thomas Henry Huxley, "by your bigness or your material resources, as such. Size is not grandeur, and territory does not make a nation. The great issue, about which hangs a true sublimity, and the terror of overhanging fate, is what are you going to do with all these things?"

The monster steam engine was an appropriate symbol of the American future, but not for the reason most of the spectators suspected. The special hopes, opportunities, and achievements, the fears and frustrations that marked the nation's grandeur in its second century—now to come—were to be even newer than visitors to the 1876 exposition could imagine. These came not from bigness but from a new kind of community. New ties would bind Americans together, would bind Americans to the larger world, and would bind the world to America. I call this community the Republic of Technology.

: 1

This community of our future was not created by any assemblage of statesmen. It had no written charter, and was not to be governed by any council of ambassadors. Yet it would reach into the daily lives of citizens on all continents. In creating and shaping this community the United States would play the leading role.

The word "Republic" I use as Thomas Paine, propagandist of the American Revolution, used it in his *Rights of Man,* to mean "not any *particular form* of government" but "the matter or object for which government ought to be instituted . . . *res-publica,* the public affairs, or the public good; or, literally translated, the public thing." This word describes the shared public concerns of people in different nations, the community of those who share these concerns.

In early modern times, learned men of the Western world considered themselves members of a Republic of Letters, the worldwide community of men who read one another's books and exchanged opinions. Long after Gutenberg's printing press had begun the process of multiplying books and encouraging the growth of literature in the languages of the marketplace, the community remained a limited one. Thomas Jefferson, for example, considered himself a citizen of that worldwide community because of what he shared with literary and scientific colleagues in France, Italy, Germany, Spain, the Netherlands, and elsewhere. When Jefferson offered the young nation his personal library (which was to be the foundation of the Library of Congress), it contained so many foreign-language books (including numerous "atheistical" works of Voltaire and other French revolutionaries) that some members of Congress opposed its purchase. The Republic of Letters was a select community of those who shared *knowledge*.

Our Republic of Technology is not only more democratic but also more in the American mode. Anyone can be a citizen. Largely a creation of American civilization in the last century, this republic offers a foretaste of American life in our next century. It is open to all, because it is a community of shared *experience*.

Behind this new kind of sharing was the Industrial Revolution, which developed in eighteenth-century England and spread over Europe and the New World. Power-driven technology and mass production meant large-scale imports and exports—goods carried everywhere in steam-driven freighters, in railroad freight cars, on transcontinental railway systems. The ways of daily life, the carriages in which people rode, the foods they ate, the pots and pans in their kitchens, the clothes they wore, the nails that held together their houses, the glass for their windows—all these and thousands of other daily trivia became more alike than they had ever been before. The weapons and tools—the rifles and pistols, the screws and wrenches, the shovels and picks—had a new uniformity, thanks to the so-called American System of Manufacturing (the system of interchangeable parts, sometimes called the Uniformity System). The telegraph and the power press and the mass-circulating newspaper brought the same information and the same images to people thousands of miles apart. Human experience for millions became more instantaneously similar than had ever been imagined possible.

This Republic of Technology has transformed our lives, adding a new relation to our fellow Americans, a new relation to the whole world. Two forces of the new era have proved especially potent.

The New Obsolescence. For most of human history, the norm had been continuity. Change was news. Daily lives were governed by tradition. The most valued works were the oldest. The great works of architecture were monuments that survived from the past. Furnishings became increasingly valuable by becoming antique. Great literature never went out of date. "Literature," Ezra Pound observed, "is news that *stays* news." The new enriched the old and was enriched by the old. Shakespeare enriched Chaucer. Shaw enriched Shakespeare. It was a world of the enduring and the durable.

The laws of our Republic of Technology are quite different. The importance of a scientific work, as the German mathematician David Hilbert once observed, can be measured by the number of previous publications it makes superfluous to read. Scientists and technologists dare not wait for their current journals. They must study "preprints" of articles and use the telephone to be sure that their work has not been made obsolete by what somebody else did this morning.

The Republic of Technology is a world of obsolescence. Our characteristic printed matter is not a deathless literary work but today's newspaper that makes yesterday's newspaper worthless. Old objects simply become second hand—to be ripe for the next season's recycling. In this world the great library is apt to seem not so much a treasure house as a cemetery. A Louis Sullivan building is torn down to make way for a parking garage. Progress seems to have become quick, sudden, and wholesale.

Most novel of all is our changed attitude toward change. Now nations seem to be distinguished not by their heritage or their stock of monuments (what was once called their civilization), but by their pace of change. Rapidly "developing" nations are those that are most speedily obsolescing their inheritance. While it took centuries or even millennia to build a civilization, the transformation of an "underdeveloped" nation can be accomplished in mere decades.

The New Convergence. The supreme law of the Republic of Technology is convergence, the tendency for everything to become more like everything else. Now the distinction is seldom made between nations that are "civilized" and those that are "uncivilized." Today, when we rely on the distinction between the "developed" and the "underdeveloped" or "developing" countries, we see the experience of all peoples converging. A common standard enables us to measure the rate of convergence statistically—by G.N.P., by per capita annual income and by rates of growth. Everyone, we assume, can participate in the newly shared experience.

A person need not be learned, or even literate, to share the fruits of technology. While the enjoyment of printed matter is restricted to those who can read, anybody can get the message from a television screen. The converging forces of everyday experience are both sublingual and translingual. People who never could have been persuaded to read Goethe will eagerly drive a Volkswagen.

The great literature that brings some people together also builds barriers. Literary classics may nourish chauvinism and create ideologies. Wars tend to reinforce national stereotypes and to harden ideologies. When the United States entered World War I, its schools ceased teaching German. Beethoven and Wagner were taboo. Still, at that very moment, American military research teams were studying German technology. Even while Indira Gandhi restricted American newsmen and American publications, she desperately tried to make the Indian technology more like the American. Technology dilutes and dissolves ideology.

In each successive world war, the competition in technology becomes more fierce—and more effective. The splitting of the atom and the exploring of space bear witness to the stimulus of competition, the convergence of efforts, the involuntary collaboration of wartime enemies. Technology is the natural foe of nationalism.

With crushing inevitability, the advance of technology brings nations together and narrows the differences between the experiences of their people. The destruction by modern warfare tends to reduce the balance of advantage between victor and vanquished. The spectacular industrial progress of Japan and Germany after World War II was actually facilitated by the wholesale destruction of their industrial plant.

Each forward step in modern technology tends to reduce the difference between the older categories of experience. Take, for example, the once elementary distinction between transportation and communication: between moving the person and moving the message. While communication once was an inferior substitute for transportation (you had to read the account because you couldn't get there), it is now often the preferred alternative. The television screen (by traditional categories a mode of communication) brings together people who still remain in their separate living rooms. With the increasing congestion of city traffic, with the parking problem, and with the lengthened holding patterns over airports, our television screen becomes a superior way of getting there. So, when it comes to public events, now you are often more there when you are here than when you are there!

Broadcasting is perhaps the most potent everyday witness to the converging powers of technology. The most democratic of all forms of public communication, broadcasting converges people, drawing them into the same experience in ways never before possible.

The democratizing impact of television has been strikingly similar to the historic impact of printing. Even in this, television's first half-century, we have seen its power to disband armies, to cashier presidents, to create a whole new democratic world—democratic in ways never before imagined, even in America. We cannot ignore the fact that the era when television became a universal engrossing American experience, the first era when Americans everywhere could witness in living color the sit-ins, the civil rights marches, was also the era of a civil rights revolution, of the popularization of protests on an unprecedented scale, of a new era for minority power, of a newly potent public intervention in foreign policy, of a new, more publicized meaning to the constitutional rights of petition, of the removal of an American President. The Vietnam War was the first American war which was a television experience. Watergate was the first national political scandal which was a television experience. The college-student protests of the sixties were the first nonsporting college events to become television experiences.

The great levelers, broadcast messages and images, go without discrimination into the homes of rich and poor, white and black, young and old. More than 96 percent of American households have at least one television set. If you own a set, no admission fee is required to enter TV land and to have a front seat at all its marvels. No questions are asked, no skill is needed. You need not even sit still or keep quiet. To enjoy what TV brings, the illiterate are just as well qualified as the educated—some would say even better qualified. Our Age of Broadcasting is a fitting climax, then, to the history of a nation whose birth certificate proclaimed that "all men are created equal" and which has aimed to bring everything to everybody.

: 2

We have reaped myriad benefits as citizens of the new Republic of Technology. Our American standard of living is a familiar name for these daily blessings. Our increased longevity, the decline of epidemics, the widening of literacy, the reduced hours of labor, the widening of political participation, our household conveniences, the reduction of the discomforts of winter and of summer, the growth of schools and colleges and universities, the flourishing of libraries and museums, unprecedented opportunities to explore the world—all are by-products of the New Obsolescence and the New Convergence. They have become so familiar that they are undervalued. But some strange fruit is apt to grow in the fertile orchards of our technological progress. If we remain aware of the special risks in the community of our future, we will run less risk of losing these unprecedented benefits that we have come to take for granted.

Here are a few of the forces at work in the Republic of Technology that will shape our American lives in the next century:

Technology invents needs and exports problems. We will be misled if we think that technology will be directed primarily to satisfying "demands" or "needs" or to solving recognized "problems." There was no "demand" for the telephone, the automobile, radio or television. It is no accident that our nation—the most advanced in technology—is also the most advanced in advertising. Technology is a way of multiplying the unnecessary. And advertising is a way of persuading us that we didn't know what we needed. Working together, technology and advertising create progress by developing the need for the unnecessary. The Republic of Technology where we will be living is a feedback world. There wants will be created not by "human nature" or by century-old yearnings, but by technology itself.

Technology creates momentum and is irreversible. Nothing can be uninvented. This tragicomic fact will dominate our lives as citizens of the Republic of Technology. While any device can be made obsolete, no device can be forgotten, or erased from the arsenal of technology. While the currents of politics and of culture can be stopped, deflected, or even reversed, technology is irreversible. In recent years, Germany, Greece, and some other countries have gone from democracy to dictatorship and back to democracy. But we cannot go back and forth between the kerosene lamp and the electric light. Our inability to uninvent will prove ever more troublesome as our technology proliferates and refines more and more unimagined, seemingly irrelevant wants. Driven by "needs" for the unnecessary, we remain impotent to conjure the needs away. Our Aladdin's lamp of technology makes myriad new genii appear, but cannot make them disappear. The automobile—despite all we have learned of its diabolism—cannot be magicked away. The most we seem able to do is to make futile efforts to appease the automobile—by building parking temples on choice urban real estate and by deferring to the automobile with pedestrian overpasses and tunnels. We drive miles—and when we are at the airport we walk miles—all for the convenience of the airplane. Our national politics is shaped more and more by the imperious demands of television. Our negotiations with the Genie of Television all seem to end in our unconditional surrender. We live, and will live, in a world of increasingly involuntary commitments.

Technology assimilates. The Republic of Technology, ruthlessly egalitarian, will accomplish what the prophets, political philosophers and revolutionaries could not. Already it assimilates times and places and peoples and things—a faithful color reproduction of the *Mona Lisa,* the voice and image of Franklin D. Roosevelt, of Winston Churchill, or of Gandhi. You too can have a ringside seat at the World Series, at Wimbledon—or anywhere else. Without a constitutional amendment or a decision of the Supreme Court, technology forces us to equalize our experience. More than ever before, the daily experience of Americans will be created equal—or at least ever more similar.

Technology insulates and isolates. While technology seems to bring us together, it does so only by making new ways of separating us from one another. The One World of Americans in the future will be a world of 250 million private compartments. The progression from the intimately jostling horse-drawn carriage to the railroad car to the encapsulated lone automobile rider and then to the seat-belted airplane passenger who cannot converse with his seatmate because they are both wearing earphones for the recorded music; the progression from the parent reading aloud to the children to the living theater with living audiences to the darkened motion-picture house to the home of private television sets, each twinkling in a different room for a different member of the family—these are the natural progressions of technology. Each of us will have his personal machine, adjusted, focused, and preselected for his private taste. CB radio now has begun to provide every citizen with his own broadcasting and receiving station. Each of us will be in danger of being suffocated by his own tastes. Moreover, these devices that enlarge our sight and vision in space seem somehow to imprison us in the present. The electronic technology that reaches out instantaneously over the continents does very little to help us cross the centuries.

Technology uproots. In this Republic of Technology the experience of the present actually uproots us and separates us from our own special time and place. For technology aims to insulate and immunize us against the peculiar chances, perils, and opportunities of our natural climate, our raw landscape. The snowmobile makes a steep mountain slope or the tongue of a glacier just another highway. Our America has been blessed by a myriad variety of landscapes. But whether we are on the mountaintop, in the desert, on shipboard, in our automobile, or on an airplane, we are protected from the climate, the soil, the sand, the snow, the water. Our roots, such as they are, grow in an antiseptic hydroponic solution. Instead of enjoying the weather given us "by Nature and by Nature's God" (in Jefferson's phrase), we worry about the humidifier and the air conditioner.

Many of these currents of change carry us further along the grand and peculiarly American course of our history. More than any other modern people we have been free of the curse of ideology, free to combine the nations, free to rise above chauvinism, free to take our clues from the delightful, unexplored, uncongested world around us. We have, for the most part, avoided the brutal homogeneities of the concentration camp and the instant orthodoxies that are revisable at the death of a Mao. During our first two centuries, a raw continent made us flexible and responsive. Our New World remains more raw and more unexplored than we will admit.

The Republic of Technology offers us the opportunity to make our nation's third century American in some novel ways.

We remain the world's laboratory. We like to try the new as do few other peoples in the world. Our experiment of binding together peoples from everywhere by opportunities rather than by ideologies will continue. The Republic of Technology offers fantastic new opportunities for opportunity.

A world where experience will be created equal tempts us in new ways and offers new dilemmas. These are the New World dilemmas of our next century. Will we be able to continue to enrich our lives with the ancient and durable treasures, to enjoy our inheritance from our nation's founders, while the winds of obsolescence blow about us and while we enjoy the delights of ever-wider sharing? Will we be able to share the exploring spirit, reach for the unknown, enjoy the multiplication of our wants, live in a world whose rhetoric is advertising, whose standard of living has become its morality—yet avoid the delusions of utopia and live a life within satisfying limits? Can we be exhilarated by the momentum that carries us willy-nilly beyond our imaginings and yet have some sense of control over our own destiny?

II Two Kinds of Revolutions

For only a tiny fragment of human history has man been aware even that he had a history. During nearly all the years since man first developed writing and civilization began, he thought of himself and of his community in ways quite different from those familiar to us today. He tended to see the passage of time, not as a series of unique, irreversible moments of change, but rather as a recurrence of *familiar* moments. The cycle of the seasons—spring, summer, fall, winter, spring—was his most vivid, most intimate signal of passing time. When men sought other useful signposts in the cycle, at first they naturally chose the phases of the moon, because the reassuring regularity of the lunar cycle, being relatively short, was easily noted. It was some time before recognition of the solar cycle (a much more sophisticated notion), with its accompanying notion of a yearly cycle, became widespread.

And, in that age of cyclical time, before the discovery of history, the repetition of the familiar provided the framework for all the most significant and dramatic occasions in human experience. Religious rituals were recreations or recapitulations of ancient original events, often the events which were supposed to have created the world. The spring was a time not only of new crops, but of a recreated earth. Just as the moon was reborn in every lunar cycle, so the year was reborn through the solar cycle.

Just as the sacred year always repeated the Creation, so every human marriage reproduced the hierogamy—the sacred union of heaven and earth. Every hero relived the career and recaptured the spirit of an earlier mythic prototype. A familiar surviving example of the age of cyclical time before the rise of historical consciousness is the Judaeo-Christian Sabbath. Our week has seven days, and by resting on the seventh day, we reenact the primordial gesture of the Lord God when on the seventh day of the Creation He "rested . . . from all his work which he had made" (Genesis 2:2).

The archaic man, as Mircea Eliade puts it, lived in a "continual present" where nothing is really new, because of his "refusal to accept himself as a historical being."

: 1

Perhaps the greatest of all historical revolutions was man's discovery—or his invention—of the idea of history. Obviously it did not occur in Western Europe on any particular day, in any particular year, or perhaps even in a particular century, but slowly and painfully. If we stop to think for a moment, we will begin to see how difficult it must have been for people whose whole world had consisted of a universe of seasons and cycles, of archetypes and resurrections, of myths relived, of heroes reincarnate, to think in a way so different.

This was nothing less than man's discovery of the new. Not of any particular sort of novelty, but of the very possibility of novelty. Men were moving from the relived-familiar, from the always-meaningful reenactment of the archetype, out into a world of unimagined, chaotic, and possibly treacherous novelty.

When did this first crucial revolution in human thought occur? In Western European civilization it seems to have come at the end of the Middle Ages, probably around the fourteenth century. The power of older ways of thinking, the dominance of cycles and rebirths, was revealed in the very name "Renaissance" (which actually did not come into use till the nineteenth century) for the age when novelty and man's power for breaking out of the cycles were discovered.

Symptoms of this new way of thinking (as Peter Burke has chronicled in his *Renaissance Sense of the Past*) are found in the writings of Petrarch (1304–1374), who himself took an interest in history, in the changing fashions in coins, clothing, words, and laws; he saw the ruins of Rome not as the creation of mythic giants but as relics of a different age. Lorenzo Valla (c. 1407–1457) pioneered historical scholarship when he proved the so-called Donation of Constantine to be a forgery, and he also laid a basis for historical linguistics when (in *De elegantia linguae latinae*) he showed the relationship between the decline of the Roman Empire and the decline of Latin. Paintings by Piero della Francesca (c. 1420–1492) and by Andrea Mantegna (1431–1506) began to abandon the reckless anachronism of earlier artists and made new efforts at historical accuracy in armor and in costume. Roman law, which would continue to dominate continental Europe, ceased to be a supra-historical, transcendental phenomenon. And other legal systems began to be seen as capable of change. In England, for example, where the common law was imagined to be the rules "to which the mind of man runneth not to the contrary," the fiction of antiquity began to dissolve, and by the seventeenth century innovation by legislation was thought to be possible. The Protestant Reformation, too, brought a new interest in historical sources and opened the way for a new kind of scrutiny of the past.

: 2

The awakening sense of history, which opened new worlds and unimagined worlds of the new, brought its own problems. Names had to be found, or made, for the particular novelties, or the kinds of novelty which history would bring. The new inquiring spirit, the newly quizzical mood for viewing the passing current of events, stirred scholars to look beneath the surface for latent causes and unconfessed motives. Early efforts to describe and explain historical change still leaned heavily on the old notion of cycles. A late version of this was offered in about 1635 in a rich baroque metaphor by Sir Thomas Browne:

> As though there were a Metempsychosis, and the soul of one man passed into another, opinions do find, after certain revolutions, men and minds like those that first begat them . . . men are lived over again, the world is now as it was in ages past . . . because the glory of one state depends upon the ruin of another, there is a revolution and vicissitude of their greatness, and must obey the swing of that wheel, not moved by intelligences [such as the souls that moved the planets] but by the hand of God, whereby all estates arise to their zenith and vertical parts according to their predestined periods. For the lives, not only of men, but of Commonwealths, and of the whole world, run not upon a helix that still enlargeth, but on a circle, where, according to their meridian, they decline on obscurity, and fall under the horizon again.

But as the historical consciousness became more lively, the historical imagination became both more sensitive and bolder.

There were more artists and scholars and lawyers and chroniclers who saw the passage of time as history.

Several words which once had a specific physical denotation began to be borrowed and given extended meanings, to describe processes in history. By the early seventeenth century (as the *Oxford English Dictionary* reveals), the word "revolution," which had described the movement of celestial bodies in an orbit or circular course and which had also come to mean the time required to complete such a full circuit, had also come into use figuratively to denote a great change or alteration in the position of affairs. In a century shaken by "commotions" (as they were sometimes called) which overthrew established governments and forcibly substituted new rulers, "revolution" came to mean what we still think of it as meaning in the twentieth century.

At about the same time, the word "progress," which until then had been used almost exclusively in the simple physical sense of an onward movement in space, and then to describe the onward movement of a story or narrative, was put to new uses. Originally neither of these senses was eulogistic. By the late seventeenth or early eighteenth century, however, "progress" had come commonly to mean advancement to a higher stage, advancement to better and better conditions, continuous improvement. That was the age of the English Enlightenment, which encompassed John Locke, Sir Isaac Newton, Robert Boyle, David Hume, and Edward Gibbon. Hardly surprising that it needed a name for progress! Similarly, by the mid-nineteenth century, as a philologist explained in 1871, the word "decadence" (derived from *de + cadere,* which meant to fall down) "came into fashion, apparently to *denote* decline and *connote* a scientific and enlightened view of that decline on the part of the user."

The century after 1776 was not only a period of great revolutions, it was also a period of great historians. In England that century produced the works of Edward Gibbon, Thomas Babington Macaulay, Henry Thomas Buckle, and W. E. H. Lecky; in the United States it was the century of Francis Parkman, William Hickling Prescott, George Bancroft, and Henry Adams. Western culture was energetically—even frantically—seeking a vocabulary to describe the new world of novelty. Historians willingly grasped at metaphors, adapted technical terms, stretched analogies, and extended the jargon of other disciplines in their quest for handles on the historical processes.

Two giants came on this scene. And—partly from the desperate need for a vocabulary, partly from their vigorous style, partly from their own towering talents for generalizing—these two have dominated much of Western writing and thinking about history into our own day. The first, of course, was Charles Darwin. In 1859 his *Origin of Species* offered with eloquent and persuasive rhetoric some strikingly new ways of describing the history of plants and animals. And he providentially satisfied the needs of man's new historical consciousness, for unlike earlier biologists he offered a way of describing and explaining the continuous emergence of novelty. Darwin brought the whole animate world into the new realm of historical consciousness. He showed that every living thing had a history. The jargon that grew out of his work, or was grafted on to his work—"evolution," "natural selection," "struggle for survival," "survival of the fittest," among other expressions—proved wonderfully attractive to historians of the human species.

There were many reasons why Darwin's vocabulary was attractive. But one of the most potent was the simplest. He provided a way of talking about change, of making plausible the emergence of novelty in experience, and of showing how the sloughing off of the old inevitably produces the new.

In Europe the nineteenth century, like the seventeenth, was an age of "commotions." After the American Revolution of 1776 and the French Revolution of 1789, revolution was in the air. And the man who translated biology into sociology, who translated the origin of species into the origin of revolutions, was Karl Marx. He freely admitted his debt to Darwin. When the first English translation of the first volume of *Das Kapital* was about to appear, Marx wrote to Darwin asking permission to dedicate the volume to him. Darwin's surprising reply was that, while he was deeply honored, he preferred that Marx not dedicate the book to him, because his family would be disturbed to have dedicated to a Darwin a book that was so Godless!

Darwin and Marx together provided the vocabulary which has dominated the writing and thinking of historians—Marxist and anti-Marxist, Communist and anti-Communist—into our own time.

Since Marx, every sort of social change has been christened a revolution. So we have the Industrial Revolution, the "Sexual Revolution," and even the so-called "Paperback Revolution." The word "revolution" has become a shorthand to amplify or dignify any subject. Revolution has become the very prototype (I could even say the stereotype) of social change.

All this reminds us that mankind has generally been more successful in describing the persisting features of his experience—warfare, state, church, school, university, corporation, community, city, family—than in describing the processes of change. Just as man has found it far simpler, when he surveys the phenomena of nature, to describe or characterize the objects—land, sea, air, lakes, oceans, mountains, deserts, valleys, bays, islands—that surround him than to describe the modes of their alteration or their motion, just as man's knowledge of anatomy has preceded his understanding of physiology, so it has been with social process.

Political changes, including the overthrow of rulers, have tended to be both more conspicuous and speedier than technological changes. Those limited numbers of people who could read and write and who kept the records have tended to be attached to the rulers and hence most aware of the changing fortunes of princes and kings.

Rapid technological change—the sort of change that can be measured in decades and that occurs within the span of a lifetime—is a characteristic of modern times. There was really no need for a name for rapid technological change until after the wave of revolutions that shook Europe beginning in mid-seventeenth century and reaching down into this century. It is during this period, of course, that men have developed their historical consciousness. The writing of history, a task of the new social sciences, only recently has become a self-conscious profession. The Regius chairs of history at Oxford and Cambridge were not established till the eighteenth century. At Harvard, the McLean professorship of history was not established till 1838, and American history did not enter the picture till much later.

: 3

What is most significant, then, about technology in modern times (the eras of most of the widely advertised "revolutions") is not so much any particular change, but rather the dramatic and newly explosive phenomenon of change itself. And American history, more perhaps than that of any other modern nation, has been marked by changes in the human condition—by novel political arrangements, novel products, novel forms of manufacturing, distribution, and consumption, novel ways of transporting and communicating. To understand ourselves and our nation, then, we must grasp these processes of change and reflect on our peculiarly American ways of viewing these processes.

In certain obvious but crucial ways, the process of technological change differs from the process of political change. I will now briefly explore these differences and suggest some of the consequences of our temptation to overlook them.

First, then, their motives (the Why). People are moved to political revolutions by their grievances (real or imagined) and by their desire for a change. Stirred by disgust with old policies and old regimes, they are awakened by visions of redress, of reforms, or of utopia. "Prudence, indeed, will dictate," Jefferson wrote in the Declaration of Independence,

> that Governments long established should not be changed for light and transient causes; and accordingly all experience hath shewn, that mankind are more disposed to suffer, while evils are sufferable, than to right themselves by abolishing the forms to which they are accustomed. But when a long train of abuses and usurpations, pursuing invariably the same Object, evinces a design to reduce them under absolute Despotism, it is their right, it is their duty, to throw off such Government, and to provide new Guards for their future security. Such has been the patient sufferance of these Colonies; and such is now the necessity which constrains them to alter their former Systems of Government.

This was a characteristically frank and clear declaration which could be a preface to most political revolutions. The Glorious Revolution of 1689 had its Declaration of Rights, the French Revolution of 1789 had its Declaration of the Rights of Man, the revolutions of 1848 had their Communist Manifesto, among others, and so it goes. For our present purpose the particular content of such declarations is less significant than that they have existed and that the people who have initiated and controlled the far-reaching political changes think of declarations as somehow giving the Why of their revolution.

But, in this sense, the great technological changes do *not* have a *Why*. The telegraph was not invented because men felt aggrieved by the need to carry messages over roads, by hand and on horseback. The wireless did not appear because men would no longer tolerate the stringing of wires to carry their messages. Television was not produced because Americans would no longer suffer the indignity or the inconvenience of leaving their homes and going to a theater to see a motion picture, or to a stadium to see a ball game. All this is obvious, but some of its significance may have escaped us. In a word, it is no trivial matter that, although in retrospect we can always see large social, economic, and geographic forces at work, still technological revolutions (by contrast with political revolutions) really have no *Why*. While political revolutions tend to be conscious and purposeful, technological revolutions are quite otherwise.

Each political revolution has its *ancien régime,* and so inevitably looks backward to what must be redressed and revised. Even if the hopes are utopian, the blueprint for utopia is made from the raw materials of the recent past. "Peace, Bread, and Land!" the slogan of the Russian Revolution of 1917, succinctly proclaimed what Russian peasants and workers had felt to be lacking. It was the obverse of "War, Starvation, and Servitude," which was taken to be a description of the *ancien régime*.

But technological revolutions generally do not take their bearings by any *ancien régime*. They more often arise not from persistent and resentful staring at the past, but from casual glimpses of what might be in the future: not so much from the pangs of empty stomachs as from the light-hearted imagining of eating quick-frozen strawberries in winter. True enough, political revolutions usually do get out of hand, and so go beyond the motives of their makers. But there usually is somebody trying to guide events to fulfill the motives of the revolutionaries, and trying to prevent events from going astray. Yet, by contrast even with the most reckless and ill-guided political revolutions, technological revolutions are still more reckless.

An example comes from World War II. From one point of view, the war in Europe was a kind of revolution, an international uprising against the Nazis, which concluded in their overthrow and removal from power. That movement had a specific objective and ran its course: surrender by the Nazis, replacement of the Nazi regime by another, "War Crimes" trials, et cetera. After that revolution took place, a Germany was left which, from a political point of view, was not radically different from pre-Nazi Germany. This was an intended result of the efforts of politicians inside and outside the country.

Now, contrast the course of what is sometimes called the Atomic Revolution, which took place during these same years. The story of the success in the United States in achieving controlled nuclear fission (which now is a well-documented chronicle) leaves no doubt that a dominant motive was the determination to develop a decisive weapon to defeat the Nazis. But the connection between Hitler and atomic fission was quite accidental. Atomic fission finally was a result of long uncoordinated efforts of scientists in many places—in Germany, Denmark, Italy, the United States, and elsewhere. And, in turn, the success in producing controlled nuclear fission and in designing a bomb spawned consequences which proved uncontrollable. Although efforts at international agreement to control the development, production, diffusion, and use of atomic weapons have not been entirely unsuccessful, the atom remains a vagrant force in the world.

The overwhelming and most conspicuous result, then, of this great advance in human technology—controlled atomic fission—was not a set of neat desired consequences. In fact, the Nazis had surrendered before the bomb was ready. Rather, as has been frequently observed, the atomic bomb was to produce vast, unpredictable, and terrifying consequences. It would give a new power to nations and level the power of nations in surprising ways. The Atomic Revolution has proved reckless, with extensive consequences and threats of consequences which make the recklessness of Hitler look like caution. Even when men think they have a *Why* for their technological revolution—as indeed Albert Einstein, Harold Urey, Leo Szilard, Enrico Fermi, and James Franck felt they had—they are deceived.

The tantalizing, exhilarating fact about great technological changes is the very fact that each such change (like the invention of controlled atomic fission) seems somehow to be a law unto itself, to have its own peculiar vagrancy. Each grand change brings into being a whole new world. But we cannot forecast what will be the rules of any particular new world until after that new world has been discovered. It can be full of all sorts of outlandish monsters; it can be ruled by a diabolic logic. Who, for example, could have predicted that the internal-combustion engine and the automobile would spawn a new world of installment buying, credit cards, franchises, and annual models—that it would revise the meaning of cities, and transform morality by instigating new institutions of no-fault reparations?

The course of political change is somehow roughly predictable, but not so in the world of technology. We discover to our horror that we are not so much masters as victims. All this is due, in part, to the wonderfully unpredictable course of human knowledge and human imagination. But it is also due (as the history of electricity, wireless communication, radio, electronics, and the transistor, among others, suggests) to all the undiscovered, still-unrevealed characteristics of the physical world. These will recreate our world and populate it with creatures we never imagined.

A second grand distinction concerns the How. It is not impossible to put together some helpful generalizations about how political revolutions are made. Some of the more familiar in modern times are those offered by Francis Bacon, Machiavelli, Montesquieu, Jefferson, John Adams, Marx, Lenin, and Mao Tse-tung. Political revolutions in modern times are the final result of long and careful planning toward specific ends, of countless clandestine meetings and numerous public rallies, of collaborative shaping toward a declared goal. Organized purposefulness, focus, clarity, and limitation of objectives—all these are crucial.

The general techniques for bringing about a political revolution—including propaganda, organization, the element of surprise, the enlistment of foreign allies, the seizure of centers of communication—have changed very little over the centuries, although of course the specific means by which these have been accomplished have changed conspicuously. John Adams, who knew a thing or two about how political revolutions were made, after the American Revolution remarked dourly on how little man had increased his knowledge of his own political processes. "In so general a refinement, or more properly a reformation of manners and improvement in science," Adams observed in 1786, "is it not unaccountable that the knowledge of the principles and construction of free governments, in which the happiness of life, and even the further progress of improvement in education and society, in knowledge and virtue, are so deeply interested, should have remained at a full stand for two or three thousand years?'' And he ventured that the principles of political science ''were as well understood at the time of the neighing of the horse of Darius as they are at this hour.'' He noted with some sadness that the ancient wisdom on these matters was still applicable.

Great changes in technology—in the very world of advancing scientific knowledge and enlarging technological grasp—paradoxically remain (as they have always been) mysterious and unpredictable. Much of the satisfaction of reading *political* history, and especially the history of political revolutions, comes from seeing men declare their large objectives, seeing them use more or less familiar techniques—and then witnessing them recognizably succeed or fail in their grand enterprise. These are the elements of frustrated ambitions and disappointed hopes, of epic and of high tragedy. But the stories of the great technological changes—even when we call them revolutions—are quite different. More often than not, it is hard to know whether an effort at technological innovation is tragedy, comedy, or bluster, whether it shows good luck or bad. How, for example, are we to assess the invention, elaboration, and universal diffusion of the airplane? Or of television?

While the patterns of political history remain in the familiar mode of Shakespeare's tragedies and historical plays (there are few changes of political regime that cannot be seen in the mold of Coriolanus, King Lear, Richard II, Richard III, Macbeth, or one of the others), technological history (despite some valiant and imaginative efforts of sociologists and historians) appears, by contrast, to have very little pattern. And much of the excitement in this story comes from the surprising coincidence, the inconceivable, and the trivial—from the boy Marconi playing with his toy, from the chance observation by a Madame Curie, from the lucky accident which befell Sir Alexander Fleming, and from myriad other occasions equally odd and unpredictable.

Even the mid-twentieth-century American Research and Development Laboratory—perhaps mankind's most highly organized, best-focused effort to promote technological change—is a place of fruitfully vagrant questing. "Research directing," explained Willis R. Whitney, the pioneer founder of the General Electric Laboratories, "is following the openings of acceptable new ideas. It is watching the growth of thought in the minds and hands of careful investigators. Even the lonely mental pioneer, being grub-staked, so to speak, advances so far into the generally unknown that a so-called director merely happily follows the new ways provided. All new paths both multiply and divide as they proceed." A modern research laboratory, then, as Irving Langmuir observed, is not so much a place where men fulfill assignments as a place where men exercise "the art of profiting from unexpected occurrences." Of course, the most adept managers of political revolutions—the Sam Adams, the Robespierre, and the Lenin—have had to know how to profit by the unexpected, but always to help them reach a prefixed destination.

The brilliant technological innovator, on the other hand, is always in search of his destination. He is on the lookout for new questions. While he hopes to find new solutions he remains alert to discover that what he thought were solutions were really new problems. Political revolutions are made by men who urge known remedies for known evils, technological revolutions by men finding unexpected answers to unimagined questions. While political change starts from problems, technological change starts from the search for problems. And, as our most adventuring scientists and technologists provide us with solutions, our society is faced with ways of preventing the newly discovered uses of the solutions (for example, the new uses of inflammable synthetics for bedclothing, nightgowns and dresses, of cellophane for packaging, of gasoline combustion for vehicles, of plastics for "disposable" containers) from themselves becoming new problems.

Of course, there are some conspicuous examples—the building of the first atomic bomb or the effort to land a man on the moon—where the purpose is specific, and where the organization resembles that of political enterprises. But here too there are special characteristics: the sense of momentum, the movement which comes from the size of the enterprise, the quantity of the investment, and the unpredictability of knowledge.

If we look back, then, on the great political revolutions and the great technological revolutions (both of which are clues to the range of mankind's capacities and possibilities) we see a striking contrast. Political revolutions, generally speaking, have revealed man's organized purposefulness, his social conscience, his sense of justice—the aggressive, assertive side of his nature. Technological change, invention, and innovation have tended, rather, to reveal his play instinct, his desire and his ability to go where he has never gone, to do what he has never done. The one shows his willingness to sacrifice in order to fulfill his plans, the other his willingness to sacrifice in order to pursue his quest. Many of the peculiar successes and special problems of our time come from our efforts to assimilate these two kinds of activities. We have tried to make government more experimental and at the same time to make technological change more purposive, more focused, more planned than ever before.

These two kinds of change—political and technological—differ not only in their Why and their How, but also in their What of It? By this I mean the special character of their consequences. Political revolutions tend, with certain obvious exceptions, to be *displacive.* The Weimar Republic displaced the regime of Imperial Germany; the Nazis displaced the Weimar Republic; and after World War II, a new republic displaced the Nazis. Normally this is what we mean by a political revolution. Moreover, to a surprising extent, political revolutions are *reversible.* In the political world, you *can* go home again. It is possible, and even common, for a new regime to go back to the ideas and institutions of an earlier regime. Many so-called revolutions are really the revivals of *anciens régimes.* The familiar phenomenon of the counter-revolution is the effort to reverse the course of change. And it is even arguable that counter-revolutions generally tend to be more successful than revolutions. The reactionary, whose objective is always more recognizable and easier to describe, thus is more apt to be successful than the revolutionary. It is the possibility of such reversals that has lent credibility to the largely fallacious pendulum theory of history, which is popularized in our day under such terms as "backlash."

Technological changes, however, thrive in a different sort of world. Momentous technological changes commonly are neither displacive nor reversible. Technological innovations, instead of displacing earlier devices, actually tend to create new roles for the devices which they might at first seem to displace. When the telephone was introduced in the later nineteenth century, some people assumed that it would make the postman obsolete (few dared predict that the United States Post Office might become decrepit before it was fully mature); similarly when wireless and then radio appeared, some wise people thought that these would spell the end of the telephone; when television came in, many were the voices lamenting the death of radio; and we still hear

Cassandras solemnly telling us that television is the death of the book. But in our own time we have had an opportunity to observe how and why such forecasts are ill founded. We have seen television (together with the automobile) provide new roles for the radio, and most recently we have seen how both have created new roles (or led to the new flourishing of older roles) for the newspaper press. And, of course, all these have created newly urgent roles for the book.

A hallmark of the great technological changes is that they tend *not* to be reversible. I have a New England friend who has not yet installed a telephone because, he says, he is waiting until it is perfected. And a few of my scholarly friends (some of them, believe it or not, eminent students, writers, and pundits about American civilization) still stubbornly refuse for even less plausible reasons to have a television set in the house. Who, having had a telephone, now does without one, or having once installed a TV set, no longer has one? There is no technological counterpart for the political restoration or the counter-revolution. Of course there are changes in style, and the antique, the obsolete, and the camp have a perennial charm. There will always, I hope, be some individuals, devotees of "voluntary simplicity," who go in search of their own Waldens. But their quixotry simply reminds us that the march of modernity is ruthless and can never retreat. In France, for example, the century following the Revolution of 1789 was an oscillation of revolutions and *anciens régimes;* aristocrats were decapitated, parties were voted out of power, old ideologies were abandoned. But during the same years the trend of technological change was unmistakable and irreversible. Unlike the French Revolution, the Industrial Revolution—despite an occasional William Morris—produced no powerful counter-revolution.

Finally, there remains a crucial difference between our abil-

ity to imagine future political revolutions and to imagine future technological revolutions. This is perhaps the most important, if least observed, distinction between the political and the technological worlds. Our failure to note this distinction I describe as the "Gamut Fallacy." "Gamut," an English word rooted in the Greek "gamma" for the lowest note in an old musical scale, means the complete range of anything. When we think, for example, of the future of our political life and our governmental forms, we can have in mind substantially the whole range of possibilities. It is this, of course, which authenticates the traditional wisdom of political theory. It illustrates what we might call "John Adams' Law" (to which I have already referred), namely, that political wisdom does not substantially progress. No wonder the astronomical analogy of "revolving" (the primary meaning of "revolution") was so tempting!

But the history of technology, again, is quite another story. We cannot envisage, or even imagine, the range of alternatives from which future technological history will be made. One of the wisest (and, surprisingly enough, one of the most cautious) of our prophets in this area is Arthur C. Clarke, the author of *2001* and other speculations. Clarke provides us with a valuable rule-of-thumb for assessing prophecies of the future of man. In his *Profiles of the Future* (after offering some instructive examples of prophecies by experts who proved beyond doubt that the atom could not be split, that supersonic transportation was physically impossible, that man could never escape from the earth's gravitational field and could certainly never reach the moon), he offers us "Arthur Clarke's Law": "When a distinguished but elderly scientist states that something is possible, he is almost certainly right. When he states that something is impossible, he is very probably wrong."

This is Clarke's way of warning us against what I have called the "Gamut Fallacy"—the mistaken notion that we can envisage all possibilities. If *any*thing is possible, then we really cannot know what is possible, simply because we cannot imagine *every*thing. Where, as in the political world, we make the possibilities ourselves, the limitations of the human imagination are reflected in the limitations of actual possibilities themselves. But the physical world is not of our making, and hence its full range of possibilities is beyond our imagining.

: 4

What are the consequences of these peculiarities of our thinking for how we can or do—or perhaps should—think about our problems today? Even in this later twentieth century, when much of mankind has begun to acquire historical consciousness, we are still plagued by the ancient problem of how to come to terms with change. The same old problem—of how to name what we so imperfectly understand, how to describe the limits of our knowledge while those very limits disqualify us from the task—still befuddles us.

Much of mankind, as we have seen, has tended to reason from the political and social to the technical, and has drawn its analogies in that direction. Faced from time immemorial with the ultimately insoluble problems of man in society, most of mankind has tended to assume that other kinds of problems might be equally insoluble. The wise prophets of the great religions have found various ways to say that, on this earth, there is no solution to the human condition. In our Western society, the parable of man's personal and social problem is the Fall of Man. "Original Sin" is another way of saying that perfection must be sought in another world, perhaps with the aid of a savior. We have been taught that in human society there are only more or less *in*soluble problems, and ultimately no solutions. The problem of politics, then, is essentially the problem of man coming to terms with his *problems*.

But our problem in the United States—and, generally speaking, the central problem of technology—is how to come to terms with *solutions*. Our misplaced hopes, our frustrations, and many of our irritations with one another and with other nations come from our unwillingness to believe in the ''insoluble'' problem, an unwillingness rooted in our New World belief in solutions. Inevitably, then, we overestimate the role of purpose in human change; we overvalue the power of wealth and the power of power.

One way of explaining, historically, how we have been tempted into this adventurous but perilous way of thinking is that we Americans have tended to take the technological problem—the soluble problem—as the prototype of the problems of our nation, and then, too, of all mankind. Among the novelties of American experience, none have been more striking than our innovations in technology, in standard of living, in the machinery of everyday life. And, as I have suggested, one of the obvious characteristics of a problem in technology is that it may really be soluble. Do you seek a way to split the atom and produce a controlled chain reaction? You have found it. That problem is solved! And so it has been with many problems, large and small, in our whole world of technology. Do you want an adhesive that will not require moistening to hold the flaps of envelopes? Do you want a highway surface that will not crack under given variations in temperature? Do you want a pen that will write under water? Do you want a camera that will produce an image in twenty seconds? Or, perhaps, do you want it in full color? We can provide you all these things. These are specific problems with specific solutions.

Taking this kind of problem as our prototype, we have too readily assumed that all other problems may be like them. While much of the rest of mankind has reasoned from the political and social to the technological (and therefore, often prematurely, drawn mistaken and discouraging conclusions), we have drawn our analogies in the other direction. By reasoning from the technological to the political and the social, we have been seduced into our own kind of mistaken, if prematurely encouraging, conclusions. It may be within our power to provide a new kind of grain and so cure starvation in some particular place. But it may not be in our power to cure injustice anywhere, even in our own country, much less in distant places.

Without being arrogant, or playing God, who alone has all solutions, we may still perhaps learn how to come to terms with our problems. We must learn, at the same time, to accept John Adams' Law (that political wisdom does not significantly progress, that the problems of society, the problems of justice and government, are not now much more soluble than they ever were, and hence the wisdom of the social past is never obsolete) while we also accept Arthur Clarke's Law (that all technological problems are substantially soluble, that "anything that is theoretically possible will be achieved in practice, no matter what the technical difficulties, if it is desired greatly enough," and hence the technological past is always becoming obsolete).

We must be willing to believe both that politics is the Art of the Possible and that technology is the Art of the Impossible. Then we must embrace and cultivate both arts. Our unprecedented American achievements both in politics and in technology therefore pose us a test, and test us with a tension, unlike that posed to any people before us in history. Never before has a people been so tempted (and with such good reason) to believe that anything is *technologically* possible. And a consequence has been that perhaps no people before us has found it so difficult to continue unabashed in search of the prudent limits of the politically possible. In this American limbo—in this new world of hope and of terror—we have a rare opportunity to profit from man's recent discovery that he has a history.

III From the Land to the Machine

When the seafaring Pilgrim Fathers disembarked from the *Mayflower* on November 21, 1620, and stepped out on their new home country, "they fell upon their knees and blessed the God of Heaven who had brought them over the vast and furious ocean, and delivered them from all the periles and miseries thereof, againe to set their feete on the firme and stable earth, their proper elemente." They were on their way to discovering, and inventing, a New World. They had committed themselves to a country that their fellow Europeans had not even imagined scarcely a century before. This might have been called the Impossible Land, for the American Continent had no place in the European's tradition. In the later Middle Ages the best authorities on the shape and extent of the known world had described a three-part planet of Europe-Asia-Africa. Their *mappae mundi* placed Jerusalem at the center and filled the rest with settled lands, real or imagined. There was no room on their map—and hardly any in their thinking, or in their history or travel literature—for a fourth continent.

By the time the Pilgrim Fathers landed, Europeans were painfully and reluctantly discovering that these shores were not part of Asia, that the Great Khan would probably not be encountered, and that the Emperor of Cipangu (Japan) would not be met on the next island. Much of what explorers had learned in the century before the Pilgrims arrived was negative. The bold settlers knew they were coming to a New World unspoiled and mostly uninhabited. But they did not yet know how new their New World might be. Despite the strenuous nostalgic efforts of several generations of colonials and "New Englanders," America would not become a New Europe.

: 1

The American experience would be different. Here men would discover new possibilities in the land, man's "proper elemente." In Europe man had shaped his notions of himself—of what he could and could not do—by his experience on familiar lands. Grandchildren and great-grandchildren usually relived their traditional experience on a friendly landscape. America offered a landscape strange and not always friendly.

There had been migrations before: the ancestors of American Indians across the Bering land bridge from Asia; the Normans into Britain, Sicily, and the Middle East; the Crusaders and their followers toward the Holy Land; the Mongols and the Turks into Eastern Europe. But most such migrations had been crusades or invasions. The ebb and flow of soldiers, nomads, bedouins, and traders had touched many lands without occupying them. The Great Atlantic Migration—in only the century and a half between 1820 and 1970—would bring some 36 million Europeans to the United States.

The American settlers came to take and shape the land. The first occupants of the land—the "Indians" whom the European migrants encountered—would not be treated, in the pattern of the Romans, as people to be incorporated into an empire. Instead, they were treated as part of the landscape. Most of them were simply cleared away, like the forests, or pushed back, like the wilderness.

By an oddity of history one large portion of the temperate regions of the planet, the heart of North America, had remained sparsely settled. When the Europeans came in the late fifteenth and sixteenth centuries, there were perhaps 2 or 3 million Indians scattered over an area about twice the size of Europe—which then had a population estimated at about 100 million. The pre-Columbian Americans had spread so thinly across North America that they had made little impression upon the land—cliff-cut pueblos in the Southwest, the tipi encampment, the occasional village. So the continent which the English and French settlers saw was almost untouched by human hand. An explorer could walk for miles through the American wilderness, or float for days down one of the broad rivers, without once seeing a sign of humankind.

Just as the Indians lacked the technology to drive off the European settlers, they also lacked the technology to change the face of the land. The land was virginal too because people elsewhere, especially Europeans, had remained so long ignorant of this part of the world. The common phrase "The Discovery of America" tells volumes about how Europeans thought—their unashamed provincialism, their isolation, the self-imprisonment of the Old World imagination.

The European encounter with the land was shaped not only by what had not happened to America but also by what had been happening in Europe. The Renaissance in Europe was an Age of Discoveries, of which the discovery of America was only one. The foundations of modern science were being laid while the Pilgrims landed at Plymouth. Francis Bacon's *Novum*

Organum persuaded men to turn from the authority of Aristotle to the evidence of their senses. Settlers who came during the seventeenth and eighteenth centuries not only possessed firearms and the knowledge to navigate thousands of miles at sea but lived in an age that was beginning to chart the flow of blood through the human body and that was tracing the planets in their orbits around the sun.

When these European settlers came to North America, there was a new kind of encounter, one that could not have happened before and which would never happen again. ''Civilized'' people—possessing the accumulated cultures of Western Europe, the inheritance of much of Arabic learning, the traditions and literatures of the classical world, the institutions, theologies, and philosophies of Judaism and Christianity, and the experience of a passage across a perilous ocean—seeking their God and their fortunes in raw and savage land. A rare opportunity!

The Puritans, who were adept at finding God's purpose in everything, explained that Divine Providence had for centuries kept this New World secret from mankind. New England, they believed, had thus been held in reserve until, at last, these English Protestants could fill it with their Puritanized religion. The Indians, then, were God's Custodians, unwittingly assigned to hold the land until the Puritans arrived.

The discovery of America did not end with the arrival of the Pilgrims. Settlers from Europe and elsewhere continued their collaborative voyages of discovery in and around and across a continent. American history, for at least a full century after the Declaration of Independence, could be summed up as a continuing discovery of America—a discovery at great cost and with great rewards—of what the land held, what people could make of the land, and how its resources could remake people's lives.

This strangely American encounter with the raw land left birthmarks on American civilization at least into the later twentieth century.

The mystery-laden faith in the future was, for much of American history, a faith in the land. The gradual unfolding of the wonders of the continent, of what could be grown on it, of what might be found under it, of how one could move up and down and across it, reinforced faith that this country was a treasure house of the unexpected. An early surprise came in the Old Northwest, the still-unmapped regions around the Great Lakes between the Ohio and the Mississippi rivers, ceded to the United States by the Treaty of Paris in 1783. This was not (as many imagined) a land of swamps and deserts but a domain of well-watered plains and fertile valleys.

And the surprises multiplied. Who could have guessed that in 1848 the streams in northern California's foothills would prove to be gold mines? Or that eleven years later there would be found in the mountains of western Nevada silver deposits rich beyond dreams of avarice? Or that the "folly" of Edwin Drake, a vagrant ex-railroad conductor, would turn out to be a treasure of flowing black mineral underneath the soil of western Pennsylvania? Who could imagine where there might be copper, coal, iron—or uranium? Who could predict where a farmer could grow sugar beets, soybeans, oranges, peanuts? Where a rancher could raise cattle for beef, sheep for wool—even alligators for luggage? Such surprising qualities of the land were not the only shaping facts of the first American centuries. But they did dominate the lives and open and define the opportunities for millions of Americans.

As the unexpected treasures of the continent-nation were revealed, as every generation uncovered some astonishing new resource, Americans quite naturally created the legend that this was a Golden Land. This legend—perhaps an overstatement but never a lie—brought more and more settlers. And Americans naturally enough believed that a God who had provided such riches for the people of His New World nation must surely have assigned them some special mission. All these once-hidden resources somehow helped persuade Americans that they had a destiny which was "manifest." Their destiny was clear, obvious, even "self-evident"—like the rights enumerated in the Declaration of Independence. Americans, then, had the further duty of discovering for all mankind all the promises still hidden in the New World.

Much of the special character of American life and American civilization, at least until the Centennial of 1876, came from the continuing encounter of post-Renaissance Europeans with pre–Iron Age America. Here was the first surprising promise of the New World, a promise that would be fulfilled in many ways. Americans would find new ways to work the land. They would build new kinds of cities—cities in a wilderness—and new kinds of schools and colleges, a new democratic world of learning. They would bring together from all over the world people with an immigrant's vision—who saw, and created, new possibilities in politics, in society, in art, in literature, in science, in technology. The promise—that civilization could transform the raw land—would explain why so many Americans were on the move, why they were so energetic at building canals, so precocious at laying railroads and at making their own kind of steamboats and locomotives. It explained the special opportunities for Americans to better their lot and rise in the world.

The rich variety of the land also helped explain why there would be a Civil War. Out of this variety would emerge problems, tragedies, and a new sense of nationhood. The Civil War, which stained with blood the first century of national life, was a conflict between opposing views of freedom, contrasting ways of life, and contrasting regions.

: 2

In the second century of national life, the land remained, and the landscapes of the continent-nation still inspired wonder. But the special qualities of American civilization were no longer the result of the encounter of sophisticated men and women with a raw continent. Now there was another, no less dramatic and no less characteristic: the encounter of Man and the Machine.

Like the other, this encounter was remarkable for its anachronism, its scale, and its speed. The new nation somehow compressed the history that Europe had experienced through two thousand years into a compact century or two. Here appeared some of the relics—slavery in the South, trial by personal combat in the West—of earlier stages of European civilization. America, though, could skip some of those stages on its way to becoming a modern nation. Moving along with unprecedented haste, America did not have to go through feudalism—with its fragmentation of loyalties, its creation of aristocracies. History here, compared to the history of Western Europe, was like a fast-motion movie, speeded up to be shown at five times the normal rate. And, in the American version, many of the episodes in the original European story were left out.

The United States never had a Middle Ages. The nation's great commercial cities—Boston, Philadelphia, Chicago, Pittsburgh—had no "city companies" or powerful, monopolizing craft guilds of the kind that had grown up over the centuries in London. In the nineteenth century this nation, by contrast with England, France, or Germany, had unexpected industrial advantages, similar to those of the bombed-out nations after World

War II. The Americans could build an industrial plant from scratch. The United States, for example, astonished the world by the pace and style of its railroad building. Railways were laid more speedily—and often more flimsily—than elsewhere, and the young United States fast outdistanced the world in railroad mileage. In Great Britain the railways grew in laborious competition with ancient roads. Foreign visitors, especially the British, marveled at how American railroads stretched "from Nowhere-in-Particular to Nowhere-at-All." This was accomplished not in spite of, but because of, the "primitiveness" of the land. In half-wild America, today's technology did not have to compete with yesterday's technology.

The United States was still only half explored when it entered the Machine Age. Even before the nation had ceased its encounter with the land, the special qualities of the machine began to put their lasting mark on American civilization. The tone and rhythm of American life—no longer the humble refrain of "Only God can make a tree"—became "Only Man can make a Machine." Americans lived in a world that every year became more man-made.

While the Machine made man feel himself master of his world, it also changed the feeling of the world that he had mastered. The Machine was a homogenizing device. The Machine tended to make everything—products, times, places, people—more alike. In the pre-Machine Age, man's life had been controlled by the weather, the landscape, the distances between places. His diet was confined by the season. In winter his house was cold; in summer it was hot. Much of what he bought was made in his neighborhood and by his neighbors. His ability to witness events was limited by the narrow range of his own eyesight. Visits to distant parts of the nation required weeks or even months, and travel was uncertain or dangerous.

The Machine changed all this. Central heating became so widespread by the mid-twentieth century that most middle-class Americans never even thought of it as an American peculiarity. Nor did they realize that central heating was a way of mastering the weather, of transforming the indoor climate from winter into summer. By the later twentieth century air conditioning completed man's mastery of the indoor climate.

Before the end of the nineteenth century the American diet had begun to be shaped by the Machine. The railroad refrigerator car brought fresh meat and milk to the cities. Canning, then refrigeration in the home, and finally quick-freezing and dehydration, made winter and summer diets more alike. By the mid-twentieth century the TV dinners that Americans ate were as unregional and as homogeneous as the network programs they watched in their living rooms. Continental distances had a new meaninglessness. The automobile had brought the city to the farmer; the airplane projected the Chicago businessman into easy reach of New York City or San Francisco. Thousands of Americans now visited Paris or Tokyo during their two-week vacations.

While this machine-mastery of the world simplified and enriched the lives of Americans in many ways, there was always a price. The golf carts that carried sedentary Americans around the courses deprived them of the pleasure of walking—and made golf a hurried, automotive sport. The snowmobile that took hordes of unskiworthy Americans across the virgin snow polluted the mountain air and shattered the mountain silence. (Perhaps the special appeal of baseball, basketball, and football was their inability to be mechanized.) Even the National Parks were not immune. This characteristic American institution became frustrated by its success. Despite the efforts of the National Park Service, some of the nation's most beautiful camping grounds were made into rural slums as cars and motorcycles brought millions to the "wilderness."

The very wonders of American democracy, which aimed to bring everything to everybody, brought new complications and confusions. Nearly everybody had more things; nearly everybody ate better, had an opportunity for more education, the chance for a better life. But were these benefits less enjoyed? Less appreciated?

The relations of Americans to their elected officials and to their governments had somehow changed. When President Thomas Jefferson received a letter, it was placed on his desk. He very likely would have opened it himself. If it merited his attention, he would have written his reply. By the middle of the twentieth century, letters directed to the President of the United States were being "processed" in the White House Mail Room, opened by an electric letter opener, and routed to one of the thousands of workers "in the White House." The few letters that reached the President's attention would get dictated replies, probably by one of the President's assistants. The letter might appear to be signed by the President. But a signature machine affixed the President's signature—or, rather, a facsimile thereof —not only to that letter but also to most of the documents that he appeared to have signed.

The factitious and the real overlapped. Not only in the White House was there a merging of the artificial and the authentic. Americans watching television were often puzzled about when and where the visible events had occurred. They wondered whether what they saw in "living color" was indeed happening then at all, whether it was "simulated" or real, fact or fiction, history or fantasy.

The Machine brought endless novelty into the world. There was hardly an activity of daily life that some device could not make more interesting—or at least more complicated. The carving knife and the toothbrush were simple tools long in use. But American inventiveness and American love of novelty would produce in time the electric knife and the electric toothbrush. And what would come next?

In the early twentieth century a philosophical American humorist, Rube Goldberg, had entertained Americans by caricaturing their love of the Machine. He also gave them an ironic motto for modern times: "Do it the hard way!" When he first began illustrating the motto in cartoons of impossible mechanisms, Americans had become newly infatuated with complicated ways of simplifying everyday life. Why walk if you could ride? Why use a wooden pencil if you could use a metal pencil with retractible lead—including many colored leads that you did not need? Or why not a ballpoint pen that could write underwater? Why write with a pencil or pen if you could use a typewriter? And why use a simple hand-operated typewriter when you could use a much more complicated electric machine? Why write it yourself at all if you could first dictate it into a machine that recorded your voice on some sort of tape, which could be put into another machine to be played back to a person who would transcribe the words on an electric typewriter? And so it went.

Just as the American's love affair with his land produced pioneering adventures and unceasing excitement in the conquest of the continent, so too his latter-day romance with the Machine produced pioneering adventures—of a new kind. There seemed to have been an end to the exploration of the landed continent—and an end to the traversing of uncharted deserts, the climbing of unscaled mountains. But there were no boundaries to a machine-made world. The New World of Machines was of man's own making. No one could predict where the boundaries might be or what his technology might make possible. To keep the Machine going, the American advanced from horse power to steam power to electrical power to internal-combustion power to nuclear power—to who could guess what.

The challenge of the Machine was as open-ended as the human spirit. Americans in the latter part of the twentieth century, in defiance of some fashionable woe-sayers, had more chance than ever before to do the unprecedented. Their problem was not the lack of opportunity for adventure but the shallowness of their human satisfaction and human fulfillment. The American challenge was how to keep alive the sense of quest which had brought the nation into being. How to discover the endless novelties of the Machine, how to make a plastic heart, devise television in three dimensions, explore the moon and planets. How to do a thousand still-unimagined works of machine magic without becoming the servant of the Machine or allowing the sense of novelty to pall or the quest for the new to lose its charm.

IV Political Technology: The Constitution

When we look back on the series of events between 1776 and 1789 which brought forth the United States of America, we must first be struck that the leaders were interested less in the ideology—the formulation of a systematic philosophy—than in the technology of politics. They were testing well-known principles by applying them to their specific problems. Their special concern was "to organize the means for satisfying needs and desires"—which is a dictionary definition of technology. There are a number of clues to this open, experimental, *technological* spirit of our North American revolutionaries.

: 1

Our first and most obvious clues are in the basic and enduring documents of the Revolution. The most important of these, of course, was the Declaration of Independence, which bore the date of July 4, 1776. The most-quoted and best-known passage, the Preamble, was actually the least characteristic. The colonists' principles were first described as "self-evident." Then "a decent Respect to the Opinions of Mankind" (as well as the exigencies of diplomacy) required a cogent summary of the causes of the particular act which they declared—the separation of thirteen British colonies. When Jefferson was accused of writing a document that had not one new idea in it, he recalled his clear, simple, practical purpose: "Not to find out new principles, or new arguments, never before thought of, not merely to say things which had never been said before; but to place before mankind the common sense of the subject, and to justify ourselves in the independent stand we are compelled to take." The body of the document applied these well-known principles—not the dogma of a particular sect, but accepted tenets of British political life during the previous century—to the conduct of the British King, who had asserted unlimited sovereignty over certain American colonists. The heart of the document was a list not of principles but of grievances. Some twenty-six items indicted the King for a wide range of specific crimes. These ranged from the King's wanton refusal to assent to needed legislation, to interference with the courts, the imposition of standing armies without the consent of the colonial legislatures, the quartering of troops on unwilling inhabitants, the protection of murderers, the obstruction of seaports, and the cutting off of trade.

Our nation's birth certificate thus unwittingly but obviously certified a congenital concern for everyday consequences. The document was not primarily a declaration of principles or a proclamation of the rights of man, it was a declaration of *independence*.

How did the Founders describe this new nation that had declared its independence so urgently? The open, empirical spirit was apparent in the very name they chose. Familiarity has dulled the meaning of the words—or rather has given them a precision they never possessed at the hour of christening. In the various documents directed to the King and Parliament during the struggle for independence this new cluster of political entities referred to itself first as "the colonies," then as "the united Colonies," and eventually as "The United Colonies of America," or "of North America." Commissions in the newly-raised army were actually written in these last two forms. The colonists' deliberative revolutionary body, when it first met in Philadelphia (September 5 to October 26, 1774) adopted for its official title nothing more explicit than "The Congress." In common usage at the time the word "Continental" was added to make "The Continental Congress," and so distinguish this one from the numerous other provincial congresses. Of course the so-called Continental Congress—representing only Atlantic seaboard colonies—was anything but continent-wide.

After the act of independence, the new nation required a name. But it was by no means clear what the country should call itself. The heading of the text of the Declaration of Independence described the enacting body as "the thirteen united [sic] states of America." The uncertainties of the situation were expressed in the fact that the word "united" (written with a lower-case "u") was treated as a mere adjective rather than as part of the proper noun. The colonists were still so dubious of their future that they dared not make "united" an indissoluble part of their nation's name.

The name finally adopted—United States of America (even later the definite article was not capitalized)—connoted all the openness that we future constituents could possibly have wished. As German Arciniegas, the brilliant Colombian man of letters has recently observed, the United States was to be the only country in the world that did not really have a name of its own. "To say *United States* is like saying *federation, republic,* or *monarchy*. Those of the north are not the only United States of America, because there exist the United States of Mexico, the United States of Venezuela, and the United States of Brazil.'' If Mexico is Mexico, he rightly noted, Venezuela is Venezuela, and Brazil is Brazil, they are all just as much a part of America as this particular North American republic. In their self-description when our North American revolutionaries chose the word ''States,'' they chose a name as indeterminate as any name that could be found for a new political entity. Incidentally, America (the word they used to locate their States) was an entity whose dimensions were only vaguely known at the time, and whose terrain (especially in North America) had barely begun to be explored. It would have been hard then to find a more imprecise geographic term. America was still a near-synonym for *terrae incognitae*.

Their final choice—United States of America—is all the more remarkable and their calculating ambiguity is all the more significant when we remember the literary talents of that generation. Believing eloquence and a feeling for poetry essential for the great statesman, in their documents and their speeches they left us many mellifluous phrases. But they gave to their greatest work, the new nation, a name that was unpoetic and even awkward, unyielding of appealing adjectives. Ambiguity acquired the look of arrogance. Now, when we citizens merely of the United States of North America arrogate to ourselves the encompassing title ''Americans,'' we still bear witness to the open and undogmatic hopes of our Founding Fathers.

While, of course, independence was what made the new nation possible, confederation was what made it durable. Despite its eloquence, the Declaration of Independence might have remained buried in colonial archives along with the early state papers of Bermuda, the Bahamas, and Jamaica if it had not been followed within a dozen years by the Constitution of the United States of America. The longevity and vitality of the Constitution came from the fact that the Framers aimed to guide the future but not fence it in. The best evidence of their self-denying intention was that their document was so brief. The Constitution of the United States, which anyone can read in an hour, is a scant 25 pages long. By contrast the constitution of my home state of Oklahoma is 158 pages, not counting amendments. Because the Framers of the federal Constitution were scrupulous to say no more than necessary, they provided a document uncannily open to the future.

With wholesome brevity came a pregnant vagueness, revealed in the very first words. The Preamble reads:

> We the People of the United States, in Order to form a more perfect Union, establish Justice, insure domestic Tranquillity, provide for the common defence, promote the general Welfare, and secure the Blessings of Liberty to ourselves and our Posterity, do ordain and establish this Constitution for the United States of America.

The three opening words ''We the People'' would prove troublesome. In their ambiguity was rooted the bloody Civil War of 1861–1865. For the leaders of the Southern States, preferring to imagine that these words really meant ''We the States,'' argued that the States which had made the Union could also dissolve it.

The Constitution was not to take effect until the people had adopted it. ''This expression [We the People],'' explained Henry (''Light-Horse Harry'') Lee, ''was introduced . . . with great propriety. This system is submitted to the people for their consideration, because on them it is to operate, if adopted. It is not binding on the people until it becomes their act.'' The Framers had the wisdom, in preparing a Constitution for posterity, not to try to elaborate or make more explicit the meaning of ''the People.'' They did not say ''we the property-owners'' or ''we the qualified voters.'' Their words, an adequate working definition in their time, would be a providential receptacle for new meanings—as civil and political rights were extended to non-property owners, to former slaves, to women, to persons above the age of eighteen, and possibly to other categories now still beyond our imagining.

All the listed purposes of the Constitution grew out of the urgencies of the Framers' recent experience. The tribulations of the loose confederation during the late war signaled the need for "a more perfect Union," the oppressive interference of the British government with the courts indicated the need to "establish Justice," recent civil disorders (Shays' Rebellion in western Massachusetts, and others elsewhere) made obvious the need to "insure domestic tranquillity," while the war itself and the later designs of European powers on the new nation showed the need to "provide for the common defence"—and so it went. This anti-doctrinaire empirical spirit would keep the document openly responsive to later needs.

: 2

If we turn from style to institutions, again we find a wholesome deference to the future. The power to amend the Constitution (Article V) was no casual item but the result of extended debate. A few members of the Constitutional Convention, led by Charles Pinckney of South Carolina, feared such a provision, for they questioned the wisdom of permitting posterity to undo their work. But George Mason countered: "The plan now to be formed will certainly be defective, as the Confederation has been found on trial to be. Amendments therefore will be necessary, and it will be better to provide for them, in an easy, regular and Constitutional way than to trust to chance and violence."

James Madison reminded the convention of the lesson to be learned from Virginia, whose "state government was the first which was made, and though its defects are evident to every person, we cannot get it amended." He drew, too, on European experience. "The Dutch have made four several attempts to amend their system without success. The few alterations made in it were by tumult and faction, and for the worse." Without some orderly means of amending the Constitution, Madison cautioned, "The fear of Innovation, and the Hue & Cry in favor of the Liberty of the people will prevent the necessary Reforms."

Finally, then, the Constitution described the means for its own amendment. The pathway that the Founders marked out for amendment would be neither facile nor impossible. There have been only twenty-six amendments. With the exception of the 18th Amendment (and its repealing 19th Amendment) concerning intoxicating beverages, the amendments have all had a constitutional dignity. Meanwhile, the very difficulty of amendment has encouraged us to exercise our ingenuity to make the original form of the Constitution workable. Our Supreme Court, filling the breach, has become a kind of continuing Constitutional Convention, reinterpreting words as circumstances require. Most important, our peaceful amending process has encouraged a continuing national debate over the needs for amendment, and has discouraged the use of violence to accomplish what is so explicitly covered by law.

The Founding Fathers not only provided (in Article V) a means to amend the Constitution, they actually provided (in Article IV) the means to amend the nation. Some doubted the wisdom of allowing the nation to expand so far that the original States might be overpowered by the new. Gouverneur Morris of New York was against allowing an unlimited number of new States to be admitted on an equality with the original thirteen.

He hoped for all time "to secure to the Atlantic States a prevalence in the National Councils." Against this provincialism the open spirit once again prevailed. James Madison and George Mason, among others, saw the promise in the unfathomed West. "The Western States," Madison insisted, "neither would nor ought to submit to a Union which degraded them from an equal rank with the other States." "If it were possible by just means to prevent emigrations to the Western Country," George Mason added, "it might be good policy. But go the people will as they find it for their interest, and the best policy is to treat them with that equality which will make them friends not enemies."

They made the process of amending the nation (by contrast with that of amending the Constitution) remarkably easy. A new State could be admitted by a simple majority vote in the Congress. The young States would in all respects be equal to their elders. Along with this came the important provision that the United States would guarantee to every State a "republican" form of government. But, after debate, the Founders wisely refused to cast this into a guarantee of any State's "existing laws." For, as William Houstoun of Georgia noted, some laws of his own State were defective, and he did not want a new federal Constitution that might become an obstacle to change. In later years, when, from time to time, the Congress tried to attach specific conditions, prohibitions, and requirements to the admission of particular States (such, for example, as were attached to the admission of Louisiana), the Supreme Court again and again declared them unconstitutional. It was this equality of States that opened the way for the United States to become a fully continental, even a transoceanic, federal republic.

In these and countless other ways the Founding Fathers declared themselves custodians of an expanding future. Federalism was their grand device for holding together experimenting communities. Each State's experiments were limited only when they violated the rights of individuals, threatened the experiments of others or weakened the whole national community. The ingenious Add-a-State Plan allowed a national laboratory to grow by installments.

"We may safely trust to the wisdom of our successors the remedies of evils to arise," Jefferson wrote to Adams less than a decade after the Constitutional Convention.

> . . . Never was a finer canvas presented to work on than our countrymen. All of them engaged in agriculture or the pursuits of honest industry, independent in their circumstances, enlightened as to their rights, and firm in their habits of order and obedience to the laws. This I hope will be the age of experiments in government, and that their basis will be founded on principles of honesty, not of mere force. We have seen no instance of this since the days of the Roman republic, nor do we read of any before that. Either force or corruption has been the principle of every modern government.

The new nation was to be not a citadel but a laboratory.

The best symbol of the Founders' experimental spirit was the federal system itself, the very framework of the new nation. In retrospect, their inspiring tentativeness stands out against the new absolute, the empyrean abstraction which others at that time imagined to be embodied in every really modern state. That abstraction was "sovereignty." It haunted governments, inflating them with an ill-founded sense of omnipotence. The feudal world of medieval Europe saw political powers, rights, and duties diffused across the land in myriad, variegated clusters. As new national states emerged after the sixteenth century, each tried to homogenize its piece of the political landscape. Each tried to build a pyramid of power, which, of course, could have only one apex.

By the later eighteenth century, British lawyers and political thinkers had imagined sovereignty to be the elixir of modern nationhood. They defined "sovereignty" as one and indivisible. "It is impossible," Governor Thomas Hutchinson of Massachusetts insisted in 1773, "there should be two independent Legislatures in the one and the same state." "In Sovereignty," Dr. Samuel Johnson wrote in *Taxation No Tyranny* (1774), "there are no gradations." For the American colonies the British saw only two alternatives, "absolute dependence" or "absolute independence."

Yet between the British government and the American colonial governments a working federalism had already emerged unannounced. While certain questions were decided in London, others were left to the capitals of the thirteen colonies. Sovereignty was diffused and divided. American federalism—a product of Atlantic distances, American space, and the slowness of communication—existed in fact long before there was an American theory. Those who ruled the British Empire remained ideologues, but American colonial leaders were glad to learn lessons from their new situation. Divided sovereignty, grown up in violation of legal metaphysics, was a leading fact of the Anglo-American experience, and a key to the American political future.

The Founding Fathers prepared the way to extend their laboratory of diffused and divided sovereignties into the full westward extent of the continent. What would happen if a growing people of varied origins and on varied landscapes went on trying federal experiments? The United States became a nation in quest of itself.

: 3

This experimental spirit, which had made the new nation politically possible, would explain much that would be distinctive of the nation's life in the following two centuries. The American limbo—a borderland between experience and idea, where old absolutes were dissolved and new opportunities discovered—would puzzle thinkers from abroad. With their time-honored distinction between fact and idea, between materialism and idealism, they labeled a people who had so little respect for absolutes as vulgar "materialists." In the gloriously filigreed cultures of the Old World it was not easy to think of life as experiment. But American life *was* experiment, and experiment was a technique for testing and revising ideas. In this American limbo all sorts of novelties might emerge. What to men of the Old World seemed a no-man's-land was the Americans' native land.

The experimentalism which had worked on the land, and would test the varied possibilities of fifty States, had found new arenas in the course of the nineteenth century. What federalism was in the world of politics, technology would be in the minutiae of everyday life. While ideology fenced in, federalism—and technology—tried out. Just as federalism would test still-unexplored possibilities in government, so technology would test unimagined possibilities in the modes of common experience.

It was not surprising that the United States would become noted—some would even say notorious—as a land of technology. The Swiss writer Max Frisch once described technology as "the knack of so arranging the world that we don't have to experience it." But in American history technology could equally well be described as "the knack of so arranging the world as to produce new experiences.'' In America the time-honored antithesis between materialism and idealism would become as obsolete as that old petrified absolute of ''sovereignty,'' which had made the British Empire come apart, and then made the American Revolution necessary. American experimentalism—in its older political form of American federalism and in its more modern generalized form of American technology—would become the leitmotif of American civilization.

V Experimenting with Education

Of all a nation's institutions, its colleges and universities—next to its churches—are the most easily petrified. In England, for example, before the end of the nineteenth century the political system had been liberalized, the franchise broadened, the economy industrialized. But Oxford and Cambridge, the centers of academic prestige and power, remained relics whose customs could be understood only by a sympathy for the Middle Ages. The Old School Tie and the college blazer remain remnants of class snobbery. Long after Americans had ceased to study Latin, and the language was employed only by medical doctors writing their prescriptions, Latin continued to be the language of college diplomas.

In view of this worldwide phenomenon of academic stasis, the story of higher education in the United States is remarkable, perhaps unique. While our colleges and universities have not failed to be citadels of the status quo, here, more than in most other nations, these institutions have been frequently and liberally irrigated by the currents of change. They have even become some of the more conspicuous areas for democratic experiment.

Needless to say, the American phenomenon has not been the

product mainly of the desire of professors to dissolve the ancient categories of their revered expertise or to enter the risky competitive marketplace. Rather it has been a by-product of characteristically American circumstances. In the United States we offer a spectacle—unfamiliar on the world scene—of the endless fluidity of the categories of knowledge, and the intimate entanglement of the so-called "higher learning" with the changing needs and desires—even the whims—of the larger community.

: 1

American education has had an odd history. In most places, and certainly in Europe, the system of education was built like a pyramid. Elementary schools prepared vast numbers of people to read and write, then smaller numbers were selected for secondary schools, and finally a tiny proportion of these were sent on to colleges and universities. This elite group at the top tended to come, of course, from the wealthy and the well-born.

Our arrangement—it should not be called a system—grew quite differently. American democracy gave a bizarre shape to our educational institutions. Instead of being a pyramid wide at the bottom, these institutions are very much like an inverted pyramid—top-heavy at the upper levels. From the traditional European point of view, our educational structure is upside down.

A clue to this oddity is the American college-founding mania which was flourishing even early in the nineteenth century. Between the beginning of the American Revolution and the end of the Civil War, less than one hundred years, more than seven hundred so-called colleges and universities were founded and had *died.* The college-founding mania continued throughout the nineteenth century, and its heyday came in the half century after the Civil War. The vast store of unoccupied land in the heart of this continent provided the opportunity for idealistic congressmen to give each State a treasure of land with which to fund its own new colleges and universities.

Jonathan Baldwin Turner, a remarkable young Yale graduate from New England who first tried to solve the Western farmers' problems by making the prickly Osage Orange hedge into a self-propagating fence, turned his missionary efforts to helping the farmer with education. His idea was to build colleges all over the West, which would be just as effective in preparing farmers for their tasks as the aristocratic Oxford and Cambridge had been effective in training Englishmen for their genteel drawing rooms, for the civil service, or for the halls of Parliament. Justin S. Morrill, a Vermont storekeeper who was sent to Congress by the new Republican party in the 1850s, had been converted to the cause of education by Turner and prepared the bill which would make possible the largest single program for higher education in modern history.

This program created the land-grant institutions. The Morrill Act of 1862, signed by President Abraham Lincoln in wartime, gave each State, from the federal public lands, a tract amounting to 30,000 acres for each of its senators and representatives in Congress. The States that did not have federal public lands within their borders were given scrip which they could use to secure public lands elsewhere. With the money from the sale of these lands each State would set up its own institutions of higher learning. The grants to States under that act totaled more than 16,000 square miles. A second Morrill Act in 1890 provided an annual federal appropriation to support the land-grant colleges, and such appropriations have increased in the present century. Religious denominations set up their own institutions. Meanwhile, men of great wealth, like Matthew Vassar, Leland Stanford, Andrew Carnegie, John D. Rockefeller, and numerous others were giving from their fortunes to found colleges and universities with the express purpose of helping to equip a democratic citizenry.

The upshot of all this was that well before the beginning of the twentieth century the United States had an astonishing array of institutions of so-called higher education. But how were Americans to be prepared to go on to these higher reaches?

The free public high school did not begin to come into existence until near the end of the nineteenth century and was itself a kind of American invention. As late as 1890 the American high school enrolled less than 7 percent of the nation's children aged fourteen to seventeen years. Of course the American system of elementary education had gone back to the colonial period and was well under way before the Civil War. But in the Old World it had been taken for granted, and so it was the widespread assumption here too, that once a person had been taught to read and write the public obligation to educate him was at an end. It was generally assumed that there was no need to make women more than literate. The relatively few academies—the preparatory schools that provided the secondary education needed to enable a person to profit from work at a college or university—were reserved for white men and for the wealthy.

The result, of course, was that Americans were trying to build the upper storeys of a democratic skyscraper without ever having built the foundations. And we see some of the effects today. One consequence was to give to colleges and universities the task of training Americans in those subjects which they should have studied in secondary school. This amounted to creating a system of high schools which bore the name and the dignity of the college. Another consequence was that the best institutions, which aimed to keep up the standards of universities, received students who were simply unprepared.

Ever since the early years of this century we have been trying to find some way of reconstructing our system of education so we can allow Americans to progress in a sensible way. Our history has not allowed us to build block on block from the bottom up. We have been desperately trying to improve both our elementary and secondary schools so that when people get to their "higher" education it really will be higher.

: 2

In the United States, by 1977, there were some 10 million students in about 3,000 institutions of higher learning. The faculties of these institutions numbered some 700,000. During most of our history, excepting certain periods of war and depression, all these figures have been steadily increasing. The "G.I. Bill" of 1944 and its successor programs (1952; 1966) offered unprecedented opportunities and inducements for veterans of World War II, the Korean War, and the Vietnam War to enter colleges and universities. During much of our recent history, the absolute numbers, the proportion of the American population in such institutions, and the rate of increase of these numbers have been significantly higher than those in other industrially developed countries. At the same time, American education (including higher education) has been characterized by the lack of any national *system*. This has, in fact, been the most important enduring feature of our education.

In place of an educational system we have had a widely diffused national program of educational experiment. Despite, even because of, this lack of a system, certain features have emerged in American education as a whole:

Community Emphasis and Community Control. American institutions of higher education have tended to be founded by communities and to be supported by communities for particular purposes. They have been expected to justify themselves to the communities which founded them (commonly defined geographically or by religious denomination). For example, Harvard College, the oldest institution of higher education in the United States, was set up in 1636 by the Massachusetts Bay Colony for a communal purpose, to provide a learned ministry. It was founded by an Act of the colony, was established with a gift from John Harvard, and then was supported by the whole colony through public appropriations and private gifts. The governing body did not consist of the scholars teaching there (as in Oxford or Cambridge colleges), but of a lay, nonacademic board which was the ancestor of all the boards of trustees that control American universities today. A continuing community emphasis has kept these American institutions under the control of community representatives, and has created and confirmed the pressure to satisfy the expectations of the community which has supported the institutions by municipal or state funds or by private donations. The spectacular growth of community colleges after World War II expressed this traditional emphasis anew and helped expand opportunities for higher education under local control.

Adaptability of Institutions and Fluidity of Subject Matters. Such institutions—founded by a particular community—have tended to be willing, or even eager, to adapt themselves to whatever at the moment has been considered to be their sponsoring community's urgent needs. Just as Harvard College aimed to provide a learned ministry for the Massachusetts Bay community, so Land-Grant institutions (many of which were originally called Agricultural and Mechanical Colleges) aimed to train farmers and their wives for rural America, and normal colleges aimed to train teachers. The host of law schools, business schools, engineering schools, schools of journalism, schools of nursing, and their descendants have aimed to provide qualified practitioners.

Traditional distinctions between high culture and low culture, between the "liberal arts" and the practical arts, and other time-hallowed distinctions have tended to dissolve. As new schools and new "programs" and projects for degrees and certificates have been freely added, the boundaries of traditional disciplines have been befogged. In England, for example, there has been a tendency to define history as what is taught or examined in the Honours School at Oxford or in the Tripos at Cambridge. But in the United States, where we have had no Oxford or Cambridge to dominate the scene, people supply their own definitions. Sometimes these are crazy, often they are faddish, but often, too, they are fertile and suggestive. New subjects enter the curriculum casually. "No Trespassing" signs are harder for professors to erect. Sociology, anthropology, psychology, economics, and statistics become more easily interfused with history, or begin to be taught in a regular curriculum. One man's sociology is another man's history.

There come to be nearly as many definitions of subjects as there are institutions; institutions compete in their definitions of subject matter and in their invention of subject matters. This fluidity has, of course, encouraged fashionable, "newsworthy," and up-to-the-minute subject matters and those which seem to have some instant vocational use. The prestige pool—both for students and faculty—is indefinitely expanded. Just as German and French officers serving with the American Revolutionary army were astounded at the omnipresence of Americans who bore the title of captain, so European visitors nowadays are understandably puzzled at the range of subjects for which Americans can be awarded the B.A. degree and at the countless American "professors."

Competition among Institutions. In countries with organized, centralized systems of higher education, there tend to be a hierarchy of institutions, a uniform salary scale, and roughly uniform conditions of employment. In the United States the rule is diversity. An instructor in one institution may receive a salary as high as that of a full professor in another; he may have a smaller teaching load and greater freedom to define his job. Institutions compete for faculty, faculty members compete for positions elsewhere. The variation in the conditions of student life, in academic standards, and in extracurricular facilities produces a widespread competition for students. The diversity can increase opportunities for self-fulfillment for both faculty and student. A student who has been disadvantaged in family or in early education can enter an easy institution and transfer to a more difficult institution with higher standards. While each institution has an incentive to be "with it" in curriculum and living conditions, and to employ the full apparatus of advertising and public relations, it also has an incentive to excel.

These characteristics of American higher education are all found in some form or other in American elementary and secondary education. Community emphasis and community control are insured by locally elected school boards. Adaptability of programs and fluidity of subject matters come from community pressures. And even the competition among institutions is expressed in the competition between parochial and public schools, between private academies and public schools, and in an increasingly mobile American population in which families with children often choose their place of residence by the character and quality of the local public schools.

: 3

All these history-rooted characteristics have been modified and confused by certain developments which have climaxed in the later twentieth-century America. These have tended to remove or reduce the benefits of our traditional experimentation and tend to substitute dogmatic central purposes—or the demands of a homogeneous populism—for the plural experimental spirit. Most of these recent developments have tended to encourage or enforce a greater uniformity in American educational institutions:

a. The interpretation of the federal Constitution, and numerous federal laws, to insure the constitutional right of students to nondiscrimination in educational opportunities. The landmark here, of course, is the desegregation decision of the Supreme Court, *Brown* v. *Board of Education* (1954). One consequence has been a general reduction in the differences between institutions, even where their differences showed a variety of interest rather than an intention to discriminate. Thus there are fewer all-male or all-female institutions.

b. Increasing sources of federal funding for education. E.g., funds for buildings, books, audio-visual aids, and numerous special programs (Head Start, etc.), the founding of, and increasing appropriations for, the National Endowments for the Arts and the Humanities.

c. Increasing federal support of scientific and technological research and development, using university faculties and facilities. An obvious example is the federal support of the research climaxing in the first nuclear chain reaction at the University of Chicago.

As much as half of the budget of some "private" institutions consists of federally funded projects. The National Institutes of Health has become a potent influence.

d. Increasing foundation support for education, research, and publication. The Ford Foundation, the Rockefeller Foundation, the Guggenheim Foundation, and a host of other foundations, large and small, operate in the national arena.

e. Increasing strength of professional organizations for teachers and specialized groups of scholars; and of accrediting organizations. For example, the American Association of University Professors (which has its rules of tenure and has blacklisted institutions); the American Federation of Teachers, and other unions. Accrediting organizations for colleges and professional schools (e.g., the North Central Association, the Association of American Law Schools, etc.) increase in power as their accreditation can affect the eligibility of an institution for sizable federal aid.

f. Increasing influence of students dominated by one or another current national political or reformist dogma.

g. Increasing pressure for sexual, racial, and other "minority" quotas for teachers and students. Often these pressures take the form of special federal and state programs, enforced by administrative or quasi-judicial bodies, and by the threats of federal agencies to withdraw federal aid.

Despite these and other pressures toward uniform standards, uniform conditions, and uniform opportunities in American educational institutions, American higher education retains many of its historic strengths and weaknesses. At best, the American situation has offered a national opportunity for creative chaos, endless variety, and open opportunity. At worst, the American situation has been anarchy and has promoted philistinism.

One notable consequence of this maelstrom has been the peculiar difficulty we Americans find in agreeing on the definition of an educated person. We become increasingly wary of traditional humanistic definitions of a liberal education, and dangerously reluctant to make literacy, much less literariness, a necessary ingredient of the highly educated.

The American experience—a federal experience with a strong tradition of community variety and local control—suggests that any effort to provide a more feasible, more precise definition of the "educated person" is not apt to succeed here through the proclamation or enforcement of national norms. Efforts to establish national standards in education have not been spectacularly successful. Where they have been somewhat effective it has been in a negative way—by finding means to prevent the violation of the rights of all citizens to equal treatment and equal opportunity. Or in the enforcement of minimum requirements (such as library facilities, numbers of Ph.D.s on the faculty, or faculty freedom from interference by boards of trustees).

The American preoccupation with the future—to which the past and present are considered only a clue—has always made it difficult here to instill a decent respect for the body of traditional learning, and the vocabulary required for that acquisition. Perhaps the closest approach to a universally acceptable American definition is Alice Freeman Palmer's "That's what education means, to be able to do what you've never done before."

VI A Laboratory of the Arts: The Immigrants' Vision

In the century after 1876, the United States became a laboratory and a symbol of the flowing together of world cultures. This convergence was a product of the daemonic energy, the focused genius, and the ambitions of numerous talented individual men and women. It was also an American by-product of the miseries bred by political totalitarianism, megalomania, and mass hysteria in far parts of the world. The United States became a museum, a workshop, and a marketplace for talents which were not tolerated elsewhere. America bore witness to the power of art and ideas to overrule legislative fiat and to overflow political boundaries.

In the perspective of American history there is an instructive irony in the impressive product of immigrant Americans during the last century. This was when, for the first time, immigration to the United States was quantitatively limited. Yet these years showed the forces of immigrant innovation to be more potent than ever before.

The product of immigrant artists in the United States demonstrated the long-run futility of the use of force to stultify or confine the acts of creation. For art obeys the opposite of Gresham's Law: quality drives out quantity. Governments may encourage population growth or birth control, they may execute, imprison, or deport individual artists or thinkers. But there is no known device for artistic contraception. The brutal tyrannies of our age have dulled and diluted the cultures of their own nations. But the *world* of culture is beyond their jurisdiction. The artists whom they discourage, punish, or expel—when these manage to escape with their lives—reappear on distant American landscapes. Here, to the freshness of their native talents they add another new dimension—their immigrants' vision.

Within this last century, such escapees and expellees helped us produce a new kind of American renaissance—a New World rebirth of Old World art and thought. Their message is peculiarly poignant because it came along with, and in spite of, a drastic change in the American spirit. In these years, despite the efforts of some of the most respectable, most "cultured" Americans, immigrant artists vindicated the American tradition of cosmopolitanism against unfriendly new American provincialisms.

: 1

The appropriate symbol of our attitude toward newcomers for the whole first century of our nation's life was the Statue of Liberty. Planned for Bedloe's Island in New York Harbor to commemorate the Centennial in 1876, it was finally unveiled by President Cleveland on October 28, 1886. On its base were inscribed Emma Lazarus' now-familiar lines:

> . . . "Give me your tired, your poor,
> Your huddled masses yearning to breathe free,
> The wretched refuse of your teeming shore.

> Send these, the homeless, tempest-tost, to me,
> I lift my lamp beside the golden door!"

Emma Lazarus spoke for the century of the Open Door.

When the Pilgrim Fathers had arrived some two hundred and fifty years before, they carried no passports (except their Bibles!) and it is doubtful how many of them could have passed an immigrant inspector's scrutiny for physical fitness and mental balance. Their opinions had a dangerously totalitarian taint. All but a small number of the forty-odd millions who followed them after Independence also arrived passportless and without "papers." They were not required to satisfy any government official of their qualifications to become Americans.

The historic American Open Door policy was, of course, a by-product of continental vastness and emptiness and remoteness. But it was not merely a historical accident. It expressed a novel principle—the American belief in the right of voluntary expatriation, the right to leave one's country and settle elsewhere. The Declaration of Independence asserted that right. As late as 1868 (when European governments were claiming jurisdiction over their nationals who had fled to the United States without their permission), the Congress declared the right of voluntary expatriation to be "a natural and inherent right of all people." The English common law had held that a subject could not change his allegiance without the permission of his government. Old World custom, reinforced by feudal institutions, had given rulers a kind of property in their peoples. Our nation, then, grew as a haven for runaways—for people who refused to endure persecution or tyranny simply because they were born under it.

But the right of voluntary expatriation was two-sided. The

right to emigrate from any place would save no one unless he also had the right elsewhere to immigrate. For the whole first century after Independence, the United States preserved both these rights substantially inviolate. The tired, the poor, the "huddled masses yearning to breathe free" not only had the right to leave the Old World, but they also were assured the right to enter the New. They poured into the United States, fully justifying Walt Whitman's boast in 1855 that we were "not merely a nation but a teeming nation of nations."

The United States adjusted itself to its immigrants and the immigrants adjusted themselves to their new country by one of two means: segregation or assimilation. Many formed their alien islands, and even hoped to preserve their isolation. The New England Puritans came here in the early seventeenth century partly because their young people had been corrupted by the laxness and heterodoxy of England or the Netherlands. Two centuries later, many who fled here from the European Revolutions of 1848 sought ways to segregate themselves. The most influential of these "Forty-Eighters" were Germans, and a considerable number of them seemed less anxious to take root in American soil than to transplant German culture. They kept the German language in their schools, they read their American German-language newspapers, they introduced "Kindergartens" and joined their own singing societies and orchestras. They came, as one contemporary put it, not to become Americanized, but to help America become Germanized.

These peculiarly American alien islands were not always formed voluntarily. Sometimes they appeared because the newcomers were socially ostracized or legally segregated. Among them were Jews, Catholics, Chinese, Africans, Mexicans, American Indians, and many others—isolated because of their "race" or supposed race, because of their unfamiliar, colorful, aggressive or passive, phlegmatic or strident ways. They fortified and solaced themselves in ethnic, racial, or religious ghettos, in neighborhoods on the wrong side of the railroad tracks, in ethnic churches and parochial schools, in lodges and brotherhoods and historical societies, in celebrations of special holidays and festivals, in spicy culinary islands of pizza parlors, kosher delicatessens, and their myriad counterparts, in defense societies and anti-defamation leagues. Their political symbol was "The Balanced Ticket."

The most important alternative to segregation was assimilation. Millions of newcomers were dissolved into the mainstream. They changed their names (or their names were changed into more pronounceable alternatives by immigration officers), they went to public schools, they intermarried with Americanized earlier immigrants, they took on the protective coloration of American speech and American clothing and an American standard of living, they joined American lodges, they converted to more "American" churches or more American sects of their Old World denomination, they became boosters for their neighborhoods, their cities and their States, they entered politics. In short, they became the justification, as they were sometimes the specific product, of movements to "Americanize" the immigrant.

: 2

At the end of the first century of our Independence, the official American attitude toward immigration was transformed. The Open Door was slammed shut—or at best left slightly ajar. The warm humanitarianism of Emma Lazarus' welcome was replaced by a wary exclusionism. The new spirit was well expressed by Thomas Bailey Aldrich's warning to the nation, in the prim *Atlantic Monthly* in 1892, to set watchmen over "The Unguarded Gates":

> Wide open and unguarded stand our gates,
> And through them presses a wild motley throng—
> Men from the Volga and the Tartar steppes,
> Featureless figures from the Hoang-Ho,
> Malayan, Scythian, Teuton, Kelt, and Slav,
> Flying the Old World's poverty and scorn;
> These bringing with them unknown gods and rites,
> Those, tiger passions, here to stretch their claws.
> In street and alley what strange tongues are these,
> Accents of menace alien to our air,
> Voices that once the Tower of Babel knew!

Though the tired immigrant's first shipboard glimpse of the promised land might still be Liberty's welcoming torch, on landing he was greeted by an unwelcoming immigration inspector.

Intellectual and social forces abroad in the land had worked this transformation. During the 1880s many ambitious young American historians and political scientists streamed to German universities. When they came back, they brought with them (along with their Ph.D.s, which became their union cards and the prototype of American graduate education) an interpretation of history which traced all good institutions—parliaments, congresses, constitutions, courts, and even the very love of liberty—to the primeval Anglo-Saxons. At the same time, the Census of 1890 reported that there was no longer a "frontier line" in the American West. This supposed "closing" of the American frontier was translated by Wisconsin historian Frederick Jackson Turner in 1893 into a frontier interpretation of American democracy. Turner's disciples nostalgically traced the

American virtues to the disappearing backwoods and to the countryside, and sounded an alarm against the crowding of American cities. President Theodore Roosevelt appointed a Commission on Country Life in 1908 to find new ways to preserve old rural values. When, in 1893, the nation suffered its worst depression till that time, the newly-unionized skilled laborers blamed their unemployment on the influx of "cheap labor" from abroad.

By 1900 these and other forces converged into the movement that closed the immigrant gates. The need to do so was rationalized by ingenious and sometimes desperate efforts to describe the prototypical American. Small groups came up with facile definitions of "Americanism."

The most potent and most respectable of these efforts was the Immigration Restriction League, founded in 1894 by three young New England blue bloods, Charles Warren, Robert DeCourcy Ward, and Prescott Farnsworth Hall. They had been persuaded in Harvard Professor Albert Bushnell Hart's History 13 that, just as Negroes had supposedly disintegrated Southern culture, so the "new" immigrants had ruined American cities. The founders of the League were joined by an impressive list of social scientists, historians, political scientists, literati, and politicians. Among them were economists Francis A. Walker, William Z. Ripley, John R. Commons, Thomas Nixon Carver, and Richard T. Ely; sociologists Franklin H. Giddings, Richard Mayo Smith, Edward A. Ross, and Robert A Woods; historians John Fiske, Albert Bushnell Hart, and Herbert Baxter Adams. The League's academic galaxy included, among others: A. Lawrence Lowell, president of Harvard; William DeWitt Hyde, president of Bowdoin College; James T. Young, director of the Wharton School of Finance; Charles F. Thwing, president of Western Reserve; Leon C. Marshall, dean of the University of

Chicago; R. E. Blackwell, president of Randolph-Macon; K. G. Matheson, president of Georgia School of Technology; and David Starr Jordan, president of Stanford. Their political spokesman was Henry Cabot Lodge.

Insisting on a crucial difference between the "old" immigration and the "new," the Immigration Restriction League and their cohorts idealized the "old" immigration into people like themselves. The Good Immigrants, whom they traced to northern and western Europe, were said to be wholesome, literate, and enterprising—eager to become Good Americans. At the same time the League caricatured the "new" immigration into a movement from eastern and southern Europe of the unskilled and the illiterate, of potential prostitutes and criminals (with an inevitable admixture of "lunatics"). These "new" immigrants, who came only because they had no alternative, would obstinately perpetuate Old World customs and values. They would never be anything but unwilling Americans.

Both the idealization and the caricature were reinforced by the conclusions of the Dillingham Commission set up by Congress in 1907 to investigate the whole problem of immigration. The commission's ponderous forty-one-volume report (1911), which included testimony and evidence of social scientists, eugenicists, economists, community leaders, and politicians, purported to mark off a historical watershed between the "old" and the "new" immigration. According to the report, those who immigrated after about 1883 had mostly come "involuntarily" (seduced by steamship and railroad advertising, and by the schemes of American employers to bring in cheap labor). The older immigrants, it was said, had helped cultivate the land, but instead the newer immigrants flooded into the cities, where they had "congregated together in sections apart from native Americans and the older immigrants to such an extent that assimilation [had] been slow.''

These ill-founded fears were being fed by news of labor troubles. The early 1870s saw the Molly Maguire riots in the Pennsylvania coal fields; the Haymarket bombing rocked Chicago in 1886; the Pullman strike, which paralyzed the railroads and brought out the federal troops, came in 1894. The radical Industrial Workers of the World (better known as the I.W.W. or the ''Wobblies'') was organized in 1904 to fight the conservative and exclusionist policies of the American Federation of Labor.

Labor troubles and other ''social disorders'' were attributed to recently-arrived immigrant ''agitators.'' When the United States entered World War I, it was said that pacifists and ''slackers'' had come mainly from this same ''foreign element,'' not real Americans but ''hyphenated'' Americans. The Bolshevik Revolution of 1917 came along to give nativists a new handle on their prejudices. ''These alien Socialists, Radicals, I.W.W.'s and Bolshevists,'' a restrictionist spokesman wrote in the *New York Times* in 1919, ''served a very useful purpose in rousing Americans to the peril of an increase in their numbers.''

Postwar prosperity in the 1920s somehow did not allay nativist fears or restrictionist passions. The Ku Klux Klan flourished anew and became a potent force in the politics of southern and midwestern States. In 1922, President A. Lawrence Lowell of Harvard (a national vice-president of the Immigration Restriction League since 1912) undertook a study of the ''race distribution'' within Harvard College. Professor Albert Bushnell Hart reported with alarm that 52 percent of the students in one course in government were ''outside the element'' from which the college had been ''chiefly recruited for three hundred years.'' Then, with his proposed quota for Jews, President Lowell aimed to prevent the ''abnormal unbalancing of races'' in American colleges.

The travail of the ''new'' immigrant was dramatized in the tragedy of Sacco and Vanzetti. Recent immigrants from Italy, these were gentle men, philosophical anarchists and pacifists, and they had avoided the draft in World War I. After they were convicted of killings at a shoe factory in Braintree, Massachusetts, Governor Alvin Fuller expressed the spirit of the times when he appointed President A. Lawrence Lowell to head the committee to review the fairness of the trial. Lowell insisted, of course, that there had been no influence of ''racial feeling'' on their trial. Sacco and Vanzetti were executed in 1927, and entered the folklore of American martyrdom, alongside Nathan Hale, John Brown, and Barbara Fritchie.

It was, of course, the descendants of earlier immigrants (in New England and the South they preferred to call their ancestors ''colonists,'' old ''settlers,'' or First Families) who led the nation into a legislative program to restrict immigration. Restrictionists showed the same kind of legalistic ingenuity that white Southern legislators had employed to disfranchise the Negro. Still, the strength of the American tradition of asylum was revealed in the deviousness of the devices to which the racial restrictionists felt themselves driven.

As early as 1897 the Immigration Restriction League, still reluctant to apply an explicitly ''racial'' standard, tried the device of a literacy test. Hoping in this way to keep out ''the undesirable classes,'' Senator Henry Cabot Lodge of Massachusetts sponsored a literacy bill which excluded any immigrant unable to read forty words in any language. It passed both houses of Congress, but was vetoed by President Cleveland, who declared that it violated the American tradition. Repeated attempts to enact a literacy bill failed. The bill which passed in 1913 was vetoed by President Taft; that which passed in 1915 was vetoed by President Wilson.

In February, 1917, on the rising wave of patriotism which preceded our entry into World War I, Congress adopted a comprehensive new Immigration Law. This incorporated a literacy test, added new classes of exclusions (chronic alcoholics, vagrants, and "persons of constitutional psychopathic inferiority"), and established a "barred zone" in the Southwest Pacific which excluded Asiatic immigrants not already kept out by the Chinese Exclusion Act of 1882 and the Gentlemen's Agreement of 1907–1908. This bill was passed over President Wilson's veto.

The next gambit of the restrictionists was a series of laws—in 1921, 1924, and 1952—which fixed an absolute number (which remained around 150,000) for total annual immigration. That number was distributed by a quota for each national group based on the proportion of people of that origin in the United States Census of some particular year (variously, 1910, 1890, or 1920). The crudity of such a device soon appeared. It was well-nigh impossible to concoct any precise definition of "national origins" for the fluid and intermixing American population. Nevertheless the facts of sociology yielded to the urgencies of politics and prejudice.

: 3

A turbulent world in the first half of the twentieth century produced hundreds of thousands of refugees. "Displaced persons"— an unhappy addition to the vocabulary of the twentieth century—described people who were not even given the opportunity to become refugees. These emerged by the thousands from Fascism, Nazism, Communism, and other forms of totalitarianism, and from the simplistic prefabricated chauvinisms of multiplying new "nations." They awakened the American conscience, and actually proved that the American tradition of the Open Door was not dead. A number of humanitarian Acts (for example, the Displaced Persons Act of 1948, the Refugee Relief Act of 1953, the Acts of 1958 to admit Hungarian political refugees, victims of earthquakes from the Azores, and Netherlands nationals from Indonesia) kept the nation's door ajar. Finally, by the Immigration Act of 1965, the "national origins" system of quotas was abandoned. But the quantitative restriction remained. After 1965 the United States cautiously returned to its Open Door tradition. The upper annual limit of 250,000 still exceeded the asylum offered by the older nations. But, by traditional American standards, the American asylum had shrunk to an un-American niggardliness.

The early decades of the twentieth century were an era when the nation's new policy of restrictive immigration was in full force. Only a few came from respectable "Anglo-Saxon" stock. Many, if not most, of the immigrant artists and intellectuals would have had to be classified in the supposedly disrespectable "new" immigration that accelerated after the 1880s. They came from "southern and eastern Europe," from Italy, Russia, Lithuania, Hungary, and Armenia—from the exotic precincts that so frightened Thomas Bailey Aldrich and his New England colleagues. Many were Jews. Most, for one reason or another, fell within classes which the restrictionists wished to exclude—and which their laws aimed to exclude.

Artists driven by the Polish and Russian pogroms, by the rise of Communism in Russia and Eastern Europe, and by the rise of

Fascism in Italy and Nazism in Germany lacked that "spontaneous" motivation which the restrictionists had idealized in their own ancestors. This was the era, par excellence, of "involuntary" immigration. People came, as D. H. Lawrence observed, not toward something, but mainly "to get away." The catastrophes they escaped were not earthquake, famine, or natural disaster. They were escaping man-made earthquakes. American civilization directly—and human civilization indirectly—would reap unpredicted benefits from Old World malevolence. And *because* these artists were displacees and refugees from new orthodoxies, from new inquisitions, from new-style pogroms, and from twentieth-century racisms, they had something special to offer.

The galaxy of artists, architects, writers, social scientists, and scientists who came in the 1930s and 1940s to escape the Nazi holocaust were the most prominent group. But they were not unique. And their characteristics typified thousands of other escapees from other holocausts. In a new sense, these were *"new* immigrants." When these men and women arrived they had already been educated in their homeland, and so arrived at the height of their achievement. They had actually been expelled because of their vigor, originality, and distinction. The United States received them full-grown, without the social cost of nurturing and training. But the economic advantage was trivial compared to another special benefit.

Those who were already fully formed and equipped—and they were thousands—could add something special to civilization here and through America to the world. They brought the most advanced and most original European modes of making and thinking for a new encounter with the American scene. Not in the minds of tourists or casual travelers—but in the persons of newly committed Americans. Each of them was a unique laboratory of the experimental spirit. They brought the immigrants' vision.

Not only in art, but in the sciences and the social sciences, the immigrant galaxy during these years was impressive. The arriving scientists and mathematicians included Albert Einstein, Max Delbrück, Leo Szilard, Enrico Fermi, and John von Neumann. Among the social scientists and psychologists were Florian Znaniecki, Hannah Arendt, Hans J. Morgenthau, Franz Alexander, Felix and Helene Deutsch, Herbert Marcuse, Karl Wittfogel, Theodor Adorno, Paul Lazarsfeld, Wolfgang Köhler, and Kurt Lewin. And these are only a sample. The catalog of composers and musicians, of art historians and publishers would show the same distinction.

Because of what they had seen, because they had been personally disavowed by their native lands, these new "new immigrants" had no desire to transplant Old World institutions or to "Europeanize" America. They enriched America, not merely like earlier immigrants with their hope and their promise, but as people who had already found their own promise, had proved their capacity for fulfillment, and welcomed new opportunities to experiment.

In no period in American history were our thought and art and culture more deeply stirred or more grandly shaped by currents from abroad. Nor had American civilization in any comparable period been more enriched by new currents. Although most of these immigrants would become "Americanized" with astonishing speed, they kept firm the remarkable individualities they had brought with them and which might not have been bred on American soil. During these very years when the United States had officially undertaken to reduce immigrant *numbers* the catalytic influence of immigrants on American culture became more powerful than ever before.

While this bore witness to the indomitable hospitality of America, which could not easily be legislated out of existence, it also bore witness to the transnational character of art and thought, to the fertility of the American soil for rebirth. Once again it proved America's capacity to be a forum and a free marketplace for the world—not merely "A Nation of Nations" but an International Nation.

VII The Fertile Machine

The beauties of the land have long been praised and sung, for the land is the proverbial source of strength. We still can see truth in the Greek myth which recounted that the giant Antaeus was invincible so long as he could touch Mother Earth. Hercules finally overcame him by lifting him into the air. Our own Thomas Jefferson put his faith in the people who live closest to the land. "Those who labour in the earth," he wrote in his *Notes on Virginia,* "are the chosen people of God, if ever he had a chosen people, whose breasts he has made his peculiar deposit for substantial and genuine virtue." Living close to the land, Jefferson added, kept a people strong and virtuous, because it kept them independent.

"Corruption of morals in the mass of cultivators," he wrote, "is a phaenomenon of which no age nor nation has furnished an example. It is the mark set on those, who not looking up to heaven, to their own soil and industry, as does the husbandman, for their subsistance, depend for it on the casualties and caprice of customers. Dependance begets subservience and venality, suffocates the germ of virtue, and prepares fit tools for the designs of ambition." Observant Americans have been impressed not merely with what people here could do with the land, but also with what intimacy with the land did to people.

Having moved into the Age of the Machine, we must take the same rounded view. We must reflect with pride and hope (if also with some caution) both on what man has done with the machine and on what the machine has done—and may do—to man.

: 1

By contrast with the land, the Machine has had a bad press. Jefferson himself expressed his strong "moral and physical preference of the agricultural, over the manufacturing, man." "It is questionable," John Stuart Mill observed, "if all the mechanical inventions yet made have lightened the day's toil of any human being."

A literary chorus declares the menace of the Machine. "Men have become the tools of their tools," warned Thoreau. Matthew Arnold declared that "Faith in machinery is . . . our besetting danger." "The world is dying of machinery," George Moore diagnosed in 1888, "that is the great disease, that is the plague that will sweep away and destroy civilisation; man will have to rise against it sooner or later." So modern a thinker as Bertrand Russell called machines "hideous, and loathed because they impose slavery." But, then, men of letters—at least until they came to live by the typewriter—were never too tolerant of innovations that enlarged the life and eased the path of the ordinary man. In the beginning, learned men had their doubts about the printing press, which would put reading matter into the hands of the great mob.

The Machine is the great witness to man's power. The land was there at the Creation. But every machine is the work of man. The power of the Machine is man's power to remake his world, to master it to his own ends. This must be a source of pride to humankind. And it may also be a source of the sin of "pride" in the special Puritan sense. It may tempt us to overlook our limitations and put ourselves in the place of God. There are some strange—even occult—features of the Machine. By inventing machines, human beings bring into the world exotic new species—tools and weapons, metal and plastic contraptions never before imagined. We have produced a chemical defoliant that outdoes any insect in its power to consume vegetation; a miraculous laser beam that excels the slicing power of any natural stone or metal and acts across vast distances; a vehicle which defecates into the atmosphere—and makes the pollution of horse dung seem trivial; a calculator that outperforms any living being in arithmetic and in applying complicated formulas.

Every inventor is a Pandora. Once mankind has created a printing press, a musket, a cotton gin, a telephone, an automobile, an airplane, a television, each of these takes on a life of its own. Biologists before Darwin mistakenly believed that no species of plant or animal could ever become extinct—because that would suggest an imperfection in God's original plan. But every machine actually has some of the qualities of an inextinguishable species. There are a few examples of machines that were forgotten for centuries before being found again. But these are rare—mere historical curiosa.

Communities, like men, find it much harder to forget than to remember. Once a machine has entered the warehouse of memory—once it has become an item of everyday use, has been described in letters, in books, and in advertisements, has been recorded in patent offices—then it requires a not-yet-invented form of magic to erase it from human experience and recollection. Even to throw it on the junk heap may prove to be a way of adding it to the future archeologist-historian's record. And, because machines commonly have been made of inorganic materials that are not readily biodegradable, the carcasses of machines remain strewn across the landscape. As our automobile cemeteries show, machines are hard to bury and not easy to cremate.

Normally when a machine enters the life of civilizations, it spawns other machines, along with novel enterprises and institutions. A machine has bizarre powers to crossbreed, to become a host, a parasite, or a saprophyte living on dead matter. Radios and air-conditioning devices find new habitats within the automobile. Enormous compacting machines come into being to give a new form to deceased automobiles—and small compacting machines arrive to make neat packages of household trash. Machines for increasing and diffusing knowledge are also fertile. The printing press made possible public schools and public libraries, along with publishers and authors who could live by their writing. The automobile brought suburbs, networks of highways, and the Drive-In Everything.

Rarely does a machine, once invented, actually disappear. It tends rather to be forgotten or to have its role transformed by another machine which does the original job more speedily, more economically, or more interestingly. The telephone was not extinguished by the radio, and radio was not extinguished by television. The daily newspaper has survived them all. The motorcycle did not obsolete the bicycle. While the automobile and the airplane seem to have won the struggle for survival against the railroads, still the railroads have proved so indispensable that we make costly efforts for their resuscitation. The lives of machines are accurately suggested in the new jargon of computers, when we speak of their first, second, or third "generation."

Bringing a new machine into the world, then, like bringing a child into the world, is a serious matter, with incalculable consequences. The power to make machines is a power to accomplish more than we can imagine, in ways we cannot predict. While tyrants and totalitarian governments may try to inhibit the inventor's imagination and can limit his resources, there has never been found an effective mode of limiting the population of ideas, nor any permanently effective machine for mind control. No pill has yet been discovered to inhibit the birth of inventions. But governments and other institutions can promote the marriages of minds and can raise the inventive birth rate.

No inventor can know an invention's precise period of gestation or the time required to bring an invention to maturity. Nor can an inventor even begin to imagine the outcome of his success. Eli Whitney surely was not trying to make a civil war. Cyrus McCormick was not intent on depopulating our farms. Henry Ford had no desire to turn choice city property into parking garages. Inventing a machine, like conceiving a child, is also motivated by personal purposes and private passions, and the act is similarly momentous and irreversible.

: 2

We have only begun to realize the Machine's occult powers. And only slowly do we discover that, however difficult it has been to govern this Nation of Nations, it may be even more difficult to govern a Nation of Machines. We have had little success in disciplining the automobile. And we are finding the airplane not much more governable. Twentieth-century American civilization, more perhaps than any other in history, is a cumulative product of innumerable passionate, thoughtless, visionary (and sometimes casual) acts of inventive conception. Our cities are a machine-made product.

Yet we have hardly begun to tell ourselves the story. We know the names of a conspicuous few inventors—the Eli Whitneys, the Cyrus McCormicks, the Alexander Graham Bells, the Henry Fords, the Thomas Edisons. These are only symbolic, just as our celebrated political and military heroes—our Adamses and Jeffersons, our Washingtons and Grants, our Lees and Eisenhowers—serve as reminders of the thousands in the ranks of citizens and soldiers.

Like those heroes, our celebrity-inventors should awaken our interest in the ranks of the common inventors who reshape our lives. Those who have most influenced everyday America—those who transformed our food, shelter, and clothing, our entertainment and information sources; those who first made a paper bag, a rotary press, a folding box, a cellophane wrapper, a picture tube, a calculating machine, or a transistor—they rarely appear in our history books.

Makers of everyday machines that remake our everyday lives have remained anonymous, partly, of course, because the inventor's work is so often collaborative, so often slowly incremental or accidental. Walter Hunt worked earnestly and with little immediate effect at inventing a sewing machine, but somehow casually in a few hours he invented the indispensable safety pin. And inventors also remain anonymous because their feats are not performed on a public platform or on the battlefield, but in attics and garages and closely guarded laboratories.

Perhaps the greatest danger in a machine-dominated America is the temptation to believe that our world is more predictable than it really is. Each triumph of our technology tempts us to redraw the geography of our imagination. We move from the world of romance and adventure into the prosaic territories of what-we-already-knew. From an open world of mystery into a world fenced by margins of error. In 1961 Isaac Asimov announced that we had "entered the age into which science-fiction authors 'escaped' a generation ago. The front pages of the newspapers read like some of the highly imaginative stories of the Thirties. The President of the United States can call for a concerted effort to place a man on the moon and be greeted with a soberly enthusiastic response. But science fiction suffers a malady no other branch of literature does. Each year sees possible plots destroyed."

The increasing quantities of technical knowledge and the growing number of specialties do threaten to fence in our imagination. What the experts saw as impossible turn out to be the spectacular technological achievements of the twentieth century—from splitting the atom to landing on the moon.

We ordinary citizens, the democratic citizenry of a technologically triumphant America—more than any other people before us—have come to take for granted everyday violations of yesterday's common sense. We accept that pictures can fly through walls and reach instantaneously over thousands of miles, that climate can be controlled, that the human heart can be repaired or replaced. In exploring the invisible atom, we do not even require as much persuading as did Ferdinand and Isabella—to invest a millionfold what they invested. We become so accustomed to men walking in the heavens that now when there is a new space performance on our television sets most of us don't even bother to watch. If we have lost some of our wholesome sense of wonder, it is wholesome too that we no longer see an opaque wall separating us from the impossible.

In our expert-ridden world, a democratic citizenry has a newly crucial role. When the expert tells us that it's impossible, *we* don't believe him! The layman's vocation is to preserve the spirit of hopeful skepticism. This is a motive to adventure, a catalyst to the imagination. "Error of opinion," Jefferson declared in his First Inaugural, "may be tolerated where reason is left free to combat it." Similarly, we need never fear the dogmatism of experts or the extravagance of our imagination so long as reason is left free to be our tonic and the marketplace of thought is kept open for the competition.

At our two hundredth birthday the nation properly emphasized its common faith: the axioms of the Declaration of Independence and the Constitution. We have, happily, shared this faith with others. In the later twentieth century the American survival of that faith is remarkable. For these two centuries have seen the greatest technological cataclysm in history and heard a chorus of the most seductive ideologies and panaceas. The more volatile, more impatient old societies say we have not been courageous but simply obstinate.

During these two centuries we have, on the whole, kept the faith proclaimed in the Declaration of Independence and the Constitution. We continue the experiment begun in the eighteenth century. We have refused to be discouraged by the most respectable nay-sayers. Never before, they tell us, has there been a nation of nations. Our strenuous quest for equal opportunity, we are told, is futile. But if we are more earnest than any nation before us to uncover our defects—and more adept at advertising them to the world—this too attests our belief that every generation of Americans must find its own ways of experiment.

We began as a Land of the Otherwise. Nothing is more distinctive, nor has made us more *un*-European than our *dis*belief in the ancient well-documented impossibilities. Every day we receive invitations to try something new. And we still give the traditional, exuberant American answer: "Why not!"

Acknowledgments

These essays bring together my recent reflections on the new meanings of technology for America and of America for technology. Two of them have never been published before; the others are revised from the versions in which they originally appeared. "The Republic of Technology" was first published in *Time,* January 17, 1977. "Two Kinds of Revolutions" is a revision of "Political Revolutions and Revolutions in Science and Technology," a lecture delivered at the National Academy of Sciences in Washington, D.C., on December 9, 1973, as part of the American Enterprise Institute's series on the Bicentennial of the American Revolution, and included in its volume *America's Continuing Revolution* (1975). "From the Land to the Machine" and "The Fertile Machine" were published in earlier form as the introductory and the concluding chapter in *We Americans* (National Geographic Society, 1976). "Political Technology: The Constitution" was delivered to the Committee on Foreign Relations of the Senate of the Republic of Brazil at Brasilia on June 22, 1976, as part of the Brazilian celebration of our Bicentennial year. "Experimenting with Education" is an expansion of my contribution to a symposium on The Educated

Person in the Modern World at the Aspen Institute for Humanistic Studies in Aspen, Colorado, July 28, 1974. "A Laboratory of the Arts: The Immigrants' Vision" is a revision of "The Immigrants' Vision," my introductory essay to *The Golden Door: Artist-Immigrants of America, 1876–1976* (Smithsonian Institution Press, 1976), the catalog of an exhibition at the Hirshhorn Museum and Sculpture Garden of the Smithsonian Institution, May 20–October 20, 1976. I wish to thank the editors of these publications for their assistance and their permission to reprint.

It has been a special pleasure in preparing this volume to have the counsel of my old friend Simon Michael Bessie, Senior Vice-President of Harper & Row.

These pages, like all my earlier publications, have been the by-product of a conversation (now for thirty-six years!), which every year becomes more delightful and more fruitful, with my wife, Ruth F. Boorstin. This volume was her idea; she has been the principal editor and has worked closely at the final revision.

Index

About the Author

Daniel J. Boorstin, the Librarian of Congress, is one of our nation's most widely read historians. His writing is known for its high literary quality as well as for its often-controversial ideas. Formerly the director of the National Museum of History and Technology of the Smithsonian Institution, Dr. Boorstin had earlier been Distinguished Service Professor at the University of Chicago, where he taught for twenty-five years. Born in Georgia and raised in Oklahoma, he received his B.A. with highest honors from Harvard and his doctor's degree from Yale. He has spent a good deal of his life viewing America from the outside, first in England, where he won a coveted "double first" while a Rhodes Scholar at Balliol College, Oxford, and was admitted as a barrister-at-law of the Inner Temple, London, and more recently as a professor living in Italy, England, France, Puerto Rico and Japan and lecturing all over the world. He is married to his editor and close collaborator, the former Ruth Frankel, with whom he has spent many hours on the mountain trails of the Rockies. The Boorstins have three grown sons.